Explaining Research

How to Reach Key Audiences to Advance Your Work

Dennis Meredith

OXFORD
UNIVERSITY PRESS

2010

Oxford University Press, Inc., publishes works that further
Oxford University's objective of excellence
in research, scholarship, and education.

Oxford New York
Auckland Cape Town Dar es Salaam Hong Kong Karachi
Kuala Lumpur Madrid Melbourne Mexico City Nairobi
New Delhi Shanghai Taipei Toronto

With offices in
Argentina Austria Brazil Chile Czech Republic France Greece
Guatemala Hungary Italy Japan Poland Portugal Singapore
South Korea Switzerland Thailand Turkey Ukraine Vietnam

Published by Oxford University Press, Inc.
198 Madison Avenue, New York, New York 10016

www.oup.com

Library of Congress Cataloging-in-Publication Data
Meredith, Dennis.
Explaining research : how to reach key audiences to advance your work / Dennis Meredith.
p. cm.
Includes bibliographical references and index.
ISBN 978-0-19-973205-0 (pbk.)
1. Communication in science. 2. Research. I. Title.
Q223.M399 2010
507.2–dc22 2009031328

9 8 7 6 5 4 3 2 1

Printed in the United States of America
on acid-free paper

Explaining Research

To my mother, Mary Gurvis Meredith.

She gave me the words.

You do not really understand something unless

you can explain it to your grandmother.

—*Albert Einstein*

PREFACE

Fortunately, I have had a multitude of heroes in my life—scientists, engineers, journalists, and fellow public information officers (PIOs). I have admired them as dedicated professionals, learned from them, and enjoyed their warm friendship. And I have deeply enjoyed writing about scientists' and engineers' discoveries, conveying those discoveries to journalists, and collaborating with my fellow PIOs.

However, throughout this gratifying career I have been acutely aware of a critical knowledge gap that I believe greatly hinders research communication. Scientists and engineers are seldom given the communication tools and techniques they need to explain their hard-won discoveries to audiences beyond their peers. And they generally do not understand journalists and PIOs well enough to work effectively with those professionals.

This guide aims to remedy that critical lack of knowledge. It distills nearly four decades of my experience as a PIO, during which I explored countless laboratories, interviewed a myriad of scientists, and prepared thousands of news releases, feature articles, Web sites, and multimedia packages.

This book aims to help you as a practicing researcher master all the tools and techniques for explaining your research—from giving compelling talks to persuading donors and administrators of the wisdom of supporting your work. Also, it aims to help you understand the journalists important to explaining your research to both lay and professional audiences. It explains the influences on their professional function and how you can work with them most effectively. In addition, a special online section at *ExplainingResearch.com* offers a guide to working with PIOs, your invaluable allies in communication. Their skills for explaining your work and reaching important audiences benefit you enormously and are invaluable to your institution.

And importantly, *Explaining Research* will show how the same tools and techniques for reaching lay audiences can greatly improve your professional communications with your colleagues.

The tools and techniques in this book can also help PIOs explain their institution's research to its many important audiences. I owe my colleagues a huge debt. I have benefited enormously from their wisdom and experience, and *Explaining Research* contains many of their ideas. Students of journalism, science, engineering, and medicine will also find this guide helpful. The communication skills it teaches will greatly benefit their future careers.

Finally, I hope this book becomes part of a continuing dialogue about the best ways to explain research to important audiences. I encourage you to visit the book's Web site, *ExplainingResearch.com*, and my blog at *ResearchExplainer.com* to take advantage of their resources and opportunities for interaction.

Although I use the term "research communication" in this book, I titled it *Explaining Research* for a reason: it covers techniques not just of clearly communicating your research, but also of *explaining* it to lay audiences that, unlike professional audiences, have no background in your field and are not inherently interested in your research. In *explaining* your work, you seek to engage and educate those audiences—to benefit your field, your institution, and your own research career.

References and Resources

The references and resources cited in this book are online at *ExplainingResearch.com*. This online listing better enables updating, enhancement, and sharing. *ExplainingResearch.com* also offers additional content such as the "Working with Public Information Officers" section to help communicate your research.

ACKNOWLEDGMENTS

My deepest thanks to the many people who gave generously of their time, their expertise, and their wisdom. They made this book immeasurably better and more insightful: Karl Bates, Sandra Blakeslee, Rick Borchelt, Chris Brodie, Merry Bruns, Robert Cooke, Keay Davidson, Tinsley Davis, Cornelia Dean, Terry Devitt, Joanna Downer, Sharon Dunwoody, Juliet Eilperin, Leslie Fink, Catherine Foster, Jon Franklin, Sharon Friedman, Lynne Friedmann, Don Gibbons, David Goldston, Chris Hildreth, Deborah Hill, Earle Holland, Michael Holland, Wendy Hunter, Deborah Illman, David Jarmul, Jim Keeley, Seema Kumar, Harvey Leifert, Jennifer Leland, Bruce Lewenstein, Alisa Machalek, Sally Maran, Stephen Maran, Maureen McConnell, Kim McDonald, Julie Miller, Steve Mirsky, Jeff Nesbit, Sue Nichols, Joe Palca, Ben Patrusky, David Perlman, Henry Petroski, Ginger Pinholster, Paul Raeburn, Rosalind Reid, Andrew Revkin, Joann Rodgers, Carol Rogers, Cristine Russell, David Salisbury, Tom Siegfried, Francis Slakey, Cathy Yarbrough, Leah Young, Patrick Young, and Bora Zivkovic.

I also thank the organizations that have made my career, and this book, possible. My professional home for some four decades has been the National Association of Science Writers, and my friends in that organization have enriched both my professional and personal life. The Council for the Advancement of Science Writing has also taught me much about science and about communication through its New Horizons in Science Briefings. For decades, that meeting has offered a savory intellectual smorgasbord of exciting science and deep insights. I also owe deep thanks to the staffs of two of the nation's leading science organizations—the American Association for the Advancement of Science and Sigma Xi, the Scientific Research Society—for their professional and personal comradeship. Their dedication to fostering communication among researchers is critical to the country's scientific and technological excellence. And, I have benefitted from working with the public affairs professionals in the leading federal research agencies—NASA, the National Institutes of Health, and the National Science Foundation. Their skills and dedication have offered invaluable lessons in how to inform the public about the discoveries made possible by public research support. I also owe a great deal to the Howard Hughes Medical Institute, for which I've long had the pleasure of writing. HHMI has provided particular inspiration because of its commitment to supporting not only research excellence but also excellence in communicating that research.

I should emphasize that, while I have drawn on the experiences and insights of all these scientists, journalists, PIOs, and institutions, this book reflects my own perspective on research communication, and any errors are entirely my own. I welcome corrections and insights that will make this book better.

Finally, I offer my deep gratitude to my wife, Joni, who has offered crucial support and unfailing enthusiasm, propping me up when mine faltered.

CONTENTS

References and resources cited in this book are listed online at *ExplainingResearch.com*.

Explaining Research

Introduction: Explaining Your Research Is a Professional Necessity

You click on the Web link or flip open the journal, and there it is: your brilliant, definitive paper! For years you conducted rigorous experiments and meticulously recorded the data. You assiduously analyzed those data to arrive at your compelling conclusions. You painstakingly wrote up the work, submitted it to a top journal, and survived a gauntlet of editors and reviewers to get it accepted. Now that it is published, your job is done…you think.

Or instead, maybe you have given your seminal talk, presenting your hard-won discoveries to your peers at a conference. You perfected your PowerPoint slides, rehearsed your delivery, and anticipated every possible question. You were eloquent, the audience was rapt, and you detected on their faces a green-with-envy tinge at your brilliance. Again, you may believe you have told the world about your research. After all, you did clearly elucidate your findings to your most important audience: your peers.

In both cases, though, if you are to do full justice to your work, your communication job has only just begun. Your paper or talk is only a first step in reaching the many audiences important to your research success: colleagues, potential collaborators in other disciplines, administrators of foundations and funding agencies, private donors, prospective students, your institution's leaders, legislators, your own family and friends, and of course the general public. Explaining your work effectively to these audiences means mounting a comprehensive communications effort—including talks, Web sites, news releases, feature articles, multimedia presentations, and media stories. Unless you take full advantage of these communication pathways, you are short-changing yourself and your research discoveries.

In fact, you are not really doing "science" unless you widely disseminate your work, argued physicist John Ziman. In his classic 1968 book *Public Knowledge: The Social Dimension of Science*, he wrote: "The objective of Science is not just to acquire information nor to utter all non-contradictory notions; its goal is a *consensus* of rational opinion over the widest possible field. [Seeking a broad consensus] is not a subsidiary consequence of the 'Scientific Method'; it is the scientific method itself." Ziman's book dealt primarily with scientific audiences. However, he would have undoubtedly agreed that the "widest possible field" includes the many lay audiences this book will help you reach.

You would not dream of switching on a new research instrument before thoroughly training yourself to use it. Nor should you try to explain your research to important lay-level audiences without learning to use communication tools and techniques. This book aims to give you those tools and techniques. And your research success depends on using these tools and techniques to explain not only a single scientific paper or talk but also your research as a whole. So, this book shows you how to fit the puzzle pieces of communication—your Web site, news releases, feature stories, and talks—into a broad strategy to portray your work to important audiences.

The communication skills this book teaches will aid your career success as well as your research success. For example, employers rank communication skills first in the qualities they seek in an applicant, according to the Job Outlook 2009 survey of more than a thousand employer organizations conducted by the National Association of Colleges and Employers. The survey found that employers ranked communication skills higher than a strong work ethic, teamwork skills, initiative, and analytical skills. As skilled as you become at communicating, you will undoubtedly encounter communication traps in explaining your work—from misleading media stories, to unfair criticism from rivals, to controversies over your findings. This book reveals those traps and shows techniques to avoid or escape them.

Lay-Level Explanations Advance Your Research

Of course, publishing excellent research papers is central to your professional success. However, lay-level explanations of your work—news releases, Web sites, videos, and so on—can convey information that even the most brilliant scientific paper cannot. For example, a scientific paper does not effectively explain the broader implications and applications of your work. It has room only for the briefest allusion to those implications. Witness perhaps the most famous such perfunctory line in the history of science—James Watson and Frances Crick's terse sentence in their 1953 *Nature* paper on the implication of their proposed structure of DNA: "It has not escaped our notice that the specific pairing we have postulated immediately suggests a possible copying mechanism for the genetic material." Needless to say, that copying mechanism provided the basis for understanding how life replicates and evolves, as well as underpinning the genomic revolution.

Lay-level articles might, in fact, be *more* effective than scientific papers at reaching some important professional audiences such as researchers in other fields. While peers in your area of expertise will read your paper, researchers outside your immediate area might not. If you are, say, a molecular biologist, you cannot expect a biomedical engineer—who could contribute ideas to your work or collaborate with you—to read the molecular biology journal that publishes your latest paper. However, that engineer might read *USA Today*, *Scientific American*, *Science*, *Nature*, *New Scientist*, or *Chemical & Engineering News*—all of which might publish articles on your research findings. Science communicator Ben Patrusky recalls many instances in which such lay-level presentations led to invaluable collaborations. For three decades, Patrusky organized the Council for the Advancement of Science Writing's New Horizons in Science Briefings for science writers, which features a variety of scientists discussing their work at a lay level. "There have been numerous collaborations developed at CASW which would not have happened but for CASW," says Patrusky. "For example, there was the geophysicist at one meeting where he heard a talk by a surgeon/geneticist about treating a critical illness. And he saw that the computational algorithm for predicting earthquakes and other catastrophes he was working on applied to that field. So, the two people—who otherwise would never have even seen one another—formed a collaboration."

Also, when posted on the Internet, lay-level communications such as news releases convey your work globally and on an equal basis with major media stories. For example, the news release on your latest paper, distributed by research news service such as EurekAlert!, will be listed on Google News and Yahoo! News right along with stories from the *New York Times* and other media outlets. In contrast, your scientific paper is far less likely to be picked up by search engines. In fact, many scientists search out news releases as quick summaries of a piece of work and its implications.

Media Coverage Affects Citations

Media coverage can also influence scientific citations of your findings by other researchers. This influence was demonstrated by a classic 1991 study in the *New England Journal of Medicine* (*NEJM*) in which David Phillips and colleagues detected an influence of newspaper coverage on scientific citations when they analyzed coverage of medical research papers in the *New York Times*. A 1978 strike at the paper gave them the comparative data they needed to correlate media coverage and citations. During that strike, reporters at *NYT* continued to select scientific papers to cover and wrote articles for "editions of record." However, these articles were not printed or distributed in published *NYT* editions.

In their analysis, the researchers compared the number of subsequent scientific citations of *NEJM* papers covered in published *NYT* articles with citations for those papers covered during the strike, but only for the record. They found that the *NEJM* papers covered in published *NYT* articles received a larger number of scientific citations than did those written during the strike. More anecdotally, my public information officer (PIO) colleagues quite commonly report that their news releases generate queries for further information from other researchers in the field, and that those queries have led to scientific contacts and to citation of the work in subsequent scientific papers.

Attention Affects Your Funding

Of course, public attention to your work will not get you a government grant; only a successful peer review of your proposal will do that. However, creating a lay-level communication plan can help that peer review. For example, in its "Broader Impacts Review Criterion" for reviewing proposals, the National Science Foundation (NSF) asks, "Will the results be disseminated broadly to enhance scientific and technological understanding?" Recognizing the power of Internet communications, NSF has broadened the definition of such disseminations beyond what most scientists understand, says Jeff Nesbit, director of NSF's Office of Legislative and Public Affairs. "Most researchers choose things that they know have worked in the past," says Nesbit. These projects usually consist of traditional educational outreach such as developing classroom materials. However, says Nesbit, "the review committees are now starting to look for the more innovative and creative ways to broaden the reach of your research, and one of the easiest ways is mass communications such as podcasts and videos."

Lay-level communications can also "raise all the research boats"—helping increase research funding by enlisting advocates for your field, says Carol Rogers,

University of Maryland journalism lecturer. "In a world where financial resources to fund research are finite, the research that is deemed to be the most significant is less likely to be on the chopping block than research that doesn't have a group of stakeholders," says Rogers. She says that "there are studies that show a correlation between visibility of research and research funding."

To be clear, greater *public* understanding of science does not necessarily lead to greater research funding, asserts Daniel Greenberg in his book *Science, Money, and Politics*. Greenberg notes that no data support a link between public understanding of science and support for science. He calls such a link "a seemingly sensible but fallacious conviction—namely that public understanding is an indispensable ingredient of public support of science." However, educating *legislators* can help increase funding, so NSF, the National Institutes of Health (NIH), and other funding agencies and advocacy groups highly prize news releases and other lay-level communications as tools in lobbying for research budgets with Congress. Chapter 19 covers how you can use your research communications most effectively to persuade legislators.

You Face a New Era of Multimedia Scientific Publication

The skills this book teaches will enhance your scientific as well as your lay-level communications. For example, you will learn to produce effective images, video, and animations for your scientific papers—necessary now that they are no longer merely "papers" but multimedia communications.

What's more, skill at broadly explaining your work serves your scientific communications, because there is a new style of scientific discourse, argues Bora Zivkovic, who is online community manager at *PLoS ONE*. This online journal exemplifies the new public, iterative style in scientific publication. It allows annotated comments, discussions, and ratings of papers by both scientists and nonscientists. Such interactivity means that researchers will need to explain their findings to broader audiences and to demonstrate the significance of those findings in the online discussion those findings evoke, says Zivkovic.

"If you write very, very dense scientese, three other people on the globe can even understand what you wrote, and they will maybe write a comment and maybe they won't," notes Zivkovic. "So you want to draw people in who maybe are not in your narrow area of expertise. You want to draw in bloggers; you want to draw educated laymen to read your paper and comment on it." Thus, he says, researchers must make the titles of their papers more broadly understandable, and they must more explicitly and clearly place their work in the context of the field.

The Web site *ResearchBlogging.org* represents a good example of the new interactive model for scientific publication. The site aggregates blog posts on peer-reviewed articles, offering science journalists and the public an independent assessment of scientific articles.

You Need to Master New Teaching Tools

If you teach, you must master the host of powerful new communication tools—Web sites, online audio and video, e-newsletters, blogs, wikis, social networks, and webinars—in order to teach your students how to use them, not to mention avoiding the embarrassment of being an outdated "technosaur" to them.

Also, by making you a better communicator, the research communication covered in this book can make you a better teacher, regardless of what you teach. Even though you may have taught many classes and given many seminars, quite likely you have had little or no formal training in pedagogical techniques. And you may not even have informal training, if you did not happen to have a mentor who was a good teacher.

Science and Engineering Lack a Culture of Explanation

This book also aims to foster a cultural change in science, engineering, and medicine—remedying their lack of an innate culture of explanation, compared to politics, sports, entertainment, and business. Certainly, you spend considerable time *communicating* to your peers—publishing scientific papers, delivering seminars, and talking shop over lunch. However, politicians, athletes, and entertainers, constantly in the public eye, are far more adept than researchers at *explaining* themselves and their work beyond their immediate colleagues. Admittedly, their fields may need less effort to explain. It is easier to talk to the public about a curve ball or a new movie than neutron stars or mitochondria. And the World Series or a movie blockbuster might at first blush appear more interesting to the public. However, I contend that science and engineering can be made just as compelling as baseball or moviemaking, and this book aims to show how.

Corporations certainly have a culture of explanation. They view communications as critical to their success, as evidenced by the vast sums they spend on advertising. Certainly, far more people can recite the latest soft drink jingle than can name even a few famous scientists, engineers, or physicians.

The miserably inaccessible form of technical papers represents another example of science and engineering's lack of a culture of explanation. Technical papers are a "communications quicksand," even though they are critical to the scientific enterprise. They often smother readers in densely packed texts of rambling, convoluted sentences.

What's more, publishers of technical journals almost willfully ignore the tenets of good design—which hold that white space, color design elements, subheads, and clear writing can aid communication. And incredibly, even online journals—where good design does not cost money in terms of paper or printing—still exhibit lousy design.

The lack of a culture of explanation shows most dramatically in how trivially scientists view lay-level communication. Witness its offhanded treatment in the 2006 book *Survival Skills for Scientists*, by Federico Rosei and Tudor Johnston. In the entire book, they devote only a single parenthetical paragraph to the topic:

> (If you are sufficiently successful in science you may be called upon to produce a popularization for the general public. At this point the only respect to be paid to the expert is to avoid saying anything actually technically incorrect, to which one can point and say, "That is clearly wrong." What you strive for in a popular presentation is (as always) clarity. Describe exactly what and how much to say. Better less and clear than more and overdense. If a technical word must be used, define it. This is all that we will say on popularization.)

I hope this book will help change this dismissive attitude—convincing you that Web sites, news releases, videos, and other communications are just as critical to your work's success as your laboratory instruments. While your instruments enable you to gather data to make discoveries, communication tools enable you to disseminate those discoveries to audiences that benefit from them and that decide about supporting your work.

Meeting the Demands of Public Science

Researchers today face more responsibilities to take a public role, and this book offers the tools to meet those responsibilities. These demands arise because the public image of science and its implications has changed drastically since federal research funding first arose in the 1950s. The public then viewed research largely as a benign activity—the source of the polio vaccine and the transistor. The huge exception, of course, was nuclear weapons, which invaded the public conscious in the form of "duck and cover" school drills and sci-fi movies featuring phalanxes of radiation-spawned giant ants, dinosaurs, and other mega-creepie-crawlies. Today, many of the important issues involving science, medicine, and engineering are highly politically charged, including global warming, stem cell research, genomic medicine, and environmental degradation.

You may well find yourself thrust unprepared into the center of public debate on such issues, says two-time Pulitzer Prize–winning journalist Jon Franklin. "They don't understand what is at stake, and they don't understand that they can't just give people the facts anymore," he says. Scientists need to understand that their communications must convey their values, not just their findings, says Franklin: "There is a need for science to be understood as a subculture. If you belong to a subculture you have to understand there are other subcultures, from accountants to the Christian right. And all these subcultures are fighting for as much ascendancy as they can get."

Your lay-level communications may benefit you in giving you the chance for the first time to explore the societal implications of your field. For example, Sharon Friedman, director of the science writing program at Lehigh University, recalls the revelations of two materials scientists when they co-taught a course on "Nanotechnology in Society" with her and a colleague who was expert in societal implications of science and technology: "They readily admitted that in their research they never think about societal implications of what they were doing, and it wasn't until they started dealing with us that they started to think about those implications," she says.

The demands of public science also create a far greater need for "citizen-scientists." These researchers recognize that their responsibilities for their field extend beyond their laboratory walls. To be a good citizen-scientist, you need not become a "public scientist"—such as physicist Michio Kaku, psychologist Steven Pinker, or astrophysicist Neil deGrasse Tyson. Rather, you need only undertake whatever communication or public service activities resonate most with your personal interests and what you believe best advances your field. For example, you might want to lobby Congress, or become a public educator, as described in chapter 18—giving a talk at your local school or taking part in programs of AAAS or your professional society.

Wielding the Power of Do-It-Yourself Communication

Dwindling media coverage of science and technology also places more responsibility on you for reaching the public directly. Coverage of science and technology occupies only a few percent of overall news coverage, according to the "State of the News Media 2008" report of the Project for Excellence in Journalism. The report found that newspapers and network TV news devote only 2 percent of their coverage to science and technology and about 7 percent to health and medicine. These percentages are far lower than for government, foreign affairs, elections and politics, crime, and economics and business.

Newspapers and magazines have drastically downsized their science and technology writing staffs in recent years. And the number of newspapers is steadily shrinking as they go out of business. So, despite the critical societal importance of science and technology, their media coverage will remain marginalized. Fortunately, the new responsibilities for explaining research also mean new opportunities, notes NSF communications officer Leslie Fink: "We still have responsibilities to the major national newspapers and the major news networks...but they're not the only players in disseminating information the way they used to be." Thus, she says, NSF has enhanced information on its Web site and launched its own news service, Science360, and other communications aimed at explaining research that it funds—just as have other funding agencies, universities, and federal laboratories.

Like these institutions, you should recognize that do-it-yourself communications can directly reach audiences that can profoundly affect your research and career success:

- **Prospective collaborators in related disciplines.** An engineer who can contribute to a research project in biology, or vice versa, might well miss even the most prominent scientific paper in the other discipline. However, broader dissemination of those results increases the likelihood that work will be communicated across disciplines.
- **Foundations and funding agencies.** Such agencies as the NSF and NIH use lay-level explanations of your research to educate legislators and the public to the importance of the work they support. They see such communications as an important part of their efforts to advocate for their budgets. And over the long term, your likelihood of getting a grant certainly increases if the agency's research budget is healthy. Visit the NSF and NIH Web sites to see their extensive lay-level coverage of research they fund. And read the special online section on working with PIOs at *ExplainingResearch.com* to learn how to work with funding agency PIOs.
- **Private donors.** Most private donors are not technically trained, so lay-level explanations of your work will help them understand the importance of their gifts.
- **Prospective students.** At universities, effective lay-level communications can attract undergraduate and graduate students to your laboratory. What's more, they will see the fact that you devote time to communications as evidence that you run an open, accessible research program.

- **Your institution's leaders.** Your department chair might understand what is going on in your lab, but your institution's trustees, president, vice presidents, and provost might not. They are often not scientifically trained. So, providing compelling, lay-level explanations of your research helps them appreciate the importance of your work to the institution. This understanding can help at budget time. More than once I have heard from researchers that they like walking into budget meetings armed with news releases and other articles that explain their work. They like even better when administrators who hold the purse strings cite such material in budget discussions and reports to trustees and donors. Institutional leaders will also appreciate your communication efforts because they enhance the reputation of the institution as a whole. In fact, when those leaders must cope with adverse media stories on the mistakes and scandals that bedevil any institution, they will appreciate even more the good news that your research discoveries represent.
- **Corporate partners.** While corporate researchers with whom you collaborate will understand the implications of your work, nontechnical executives may not. And they are the ones who ultimately approve and advocate for those collaborations.
- **Legislators.** News releases and feature articles help your institution's government relations officers make the case with state and national legislators to support your institution and its work. At one time or another, you may find yourself visiting those legislators, or even testifying before a legislative committee. It helps greatly in those encounters if legislators and their staff can be prepped with clear, accessible explanations of your work.
- **Your own family and friends.** So many times I have heard from researchers, commenting on news releases and features, that "at last my family and friends will understand what I do!" Such understanding might make it a little easier for spouses to understand a researcher's long hours in the lab, or may have helped researchers through a bit of awkwardness at family reunions by helping the family understand how cool the research is.

This book will teach you how to become an adept do-it-yourself research communicator.

Explaining Your Work Protects You Professionally

News releases, feature articles, and other communications protect you in important ways. They constitute your approved public statement about your research and its implications, explained precisely how you want it explained and giving credit to colleagues and funding agencies.

Such public statements can be critically important because invariably some media reports on your work will misrepresent your experiments, fail to give credit to colleagues, or misconstrue its implications. In such cases, you can point to your own lay-level accounts of your work as the authoritative source of information on your work. What's more, your news releases offer an instant antidote for mistakes, because they appear alongside those media reports on Google, Yahoo!, and other search engines and Web news sites. Thus, for example, if a collaborator feels slighted by a newspaper or magazine report, your release is proof that your public statement does give full credit and that you are not out to grab all the glory.

Also, if your work could be misconstrued by the media, you can preempt that possibility with well-crafted lay-level communications. For example, Duke neurobiologist Michael Platt published a paper showing that monkeys would rather glimpse photos of female hindquarters than receive a juice reward. Media stories or blogs might have made fun of the work, missing its real scientific significance. So, the news release I wrote led with the fact that the research demonstrated a valuable animal model for studying autism, since the method enabled precise measurements of primate social sense. Invariably, some news stories took a humorous slant, with headlines saying that "Monkeys Like Porn." However, most of these stories also included the significance of the work in understanding social sense and how it might malfunction in autism. And, the news release was posted on the university Web site, Google, and Yahoo! as an antidote to such misinterpretation.

Finally, if you do not choose to proactively explain your research, by default you leave such communications to people not as familiar with your work and to the informal grapevine. Your research might well be explained for you in an uninformed way over which you have no control or influence.

Of course, your scientific papers and proposals will contain the precise, technical descriptions of your work. But those communications are not as accessible and, in fact, not interesting to the many audiences beyond your colleagues that you want to understand and appreciate your work.

But Will You Be Pegged as a Publicity Hound?

In the olden days, scientists who sought publicity for their research were sometimes accused by their colleagues of being publicity-hungry self-promoters. Such worries reflect 20th-century thinking. Today, the great majority of your fellow researchers and your institution's administrators are savvy enough to understand how important it is to explain your work to the key audiences listed above. Most likely, the people who criticize your communication efforts will be either those whose research is not significant enough to warrant such communications, or those who are naive about the value of research communications. Such criticisms also tend to be merely vague grumblings, rather than substantive comments, and certainly not significant enough to compromise your scientific career. The benefits of explaining your research responsibly vastly outweigh any such sniping, and you should ignore it.

However, those "public scientists" such as Carl Sagan, who assumed the role of a popular educator about science, have suffered for their public role; and chapter 27 explores the pros and cons of being a public scientist.

Will Explaining Your Work Detract from Your Professional Duties?

Even given the extraordinary value of lay-level research communications, you may worry that the effort will take too much valuable time from your research and other professional duties. However, lay-level communications contribute significantly to your ability to carry out your professional duties. For one thing, they give you a chance to hone the same skills and techniques that you will apply to communicating with your colleagues. Giving talks, making news videos, and helping with news releases will make your professional communications immeasurably better.

True, as you scan the table of contents of this book, you may feel a bit overwhelmed by the multitude of ways to communicate your work—talks, news releases, Web sites, videos, blogs, podcasts, webinars, and so on. You may feel that the time investment is just too much. However, you will be expected to create many of these communications anyway, so why not invest a bit of effort in making them professional quality and more effective?

Also, recognizing how precious your time is, this book will show you how to develop a "strategy of synergy" that enables you to plan and carry out your communications to make the benefits greatly outweigh the costs in time and energy. What's more, the special online section at *ExplainingResearch.com* on working with PIOs will show you how to enlist their services to help your communications. These services can include

- Writing and distributing news releases
- Creating photos and multimedia packages

- Providing clippings
- Developing media and communication strategies
- Briefing and scouting journalists
- Giving you media credibility
- Managing crisis communications

Finally, of course, you have a duty to explain your research to the society that supports your work. As then-AAAS president John Holdren told attendees at the 2007 annual meeting:

> Scientists and technologists need to improve their communication skills, so that they can convey the relevant essence of their understandings to members of the public and to policymakers. They need to seek out avenues for doing that. And I believe that every scientist and technologist should tithe ten percent of his or her professional time and effort to working to increase the benefits of science and technology for the human condition and to decrease the liabilities. The challenges demand no less.

PART I

Learning a New Communications Paradigm

1

Understand Your Audiences

Although you have more sense than to spout equations in public debates like the scientists in the cartoon that introduces this section, like them you probably have not thought much about what's inside the heads of the lay audiences you address. Because you spend most of your time talking to and writing for your peers, you might have settled into the comfortable rut of perceiving lay audiences as a homogeneous lot.

You think that lay audiences, like your peers, all understand your technical terms and concepts, pay uniformly rapt attention to your talks, and read your papers with great concentration. However, the audiences for lay-level communications differ from your professional audience and from one another. They differ in what they need in both the substance and style of your research explanations. Accept that you will have to work harder and smarter to reach lay audiences. However, as a bonus, as you master the lay communications techniques this book teaches, you will also reach your professional audiences more effectively.

Besides being different from one another, both lay and professional audiences are more internally heterogeneous than you may understand. A lay audience might range from people who have never set foot in a science classroom to those with advanced science degrees. A professional audience might range from undergraduates new to your field to senior scientists who know every conceptual nook and cranny. What's more, each audience member has "multiple personalities," says University of Maryland journalism lecturer Carol Rogers: "We people sitting in an audience may be scientists, but we also may be parents, children, donors, or activists. We may come at a topic from different cultural perspectives. Keeping that in mind could be unnerving, but it can also just give you a richer appreciation for the many levels of interaction and understanding that may be going on."

They See You as a Hero

Fortunately, you start with the advantage that lay audiences see you as a hero. The vast majority of scientists and engineers in movies and television shows are portrayed as heroes, from archeologist Indiana Jones to television's *CSI* forensic scientists. In movies and television shows, scientists and engineers usually save the world or solve a crime. If you do not believe me, explore the extensive lists of Hollywood hero and villain scientists at *ExplainingResearch.com*—evil scientists are relatively few and far between. And many times, they are not so much evil as misguided, ultimately seeing the error of their ways.

Besides scientists' Hollywood hero status, the public trusts them even more than the reporters who write about them. In a 2006 Harris Poll, Americans said they trusted doctors (85 percent), teachers (83 percent), scientists (77 percent), and professors (75 percent) far more than they did journalists (39 percent), lawyers (27 percent), or pollsters (34 percent). What's more, polls rank scientists and engineers high in contributing to society. Respondents in a 2009 survey by the Pew Research Center for People and the Press said that people contributing the most to society's well-being were members of the military, teachers, scientists, medical doctors, and engineers. And according to the National Science Board's *Science and Engineering Indicators 2008*, "more Americans expressed a great deal of confidence in leaders of the scientific community than in the leaders of any other institution except the military." This trustworthiness gives you the credibility to define how you and your research are portrayed to your audiences. It means your news releases, Web sites, videos, and so forth, are just as credible as media stories.

They Like Science but Have Reservations

Your audiences are also already predisposed to like science and be interested in it. For example, the 2009 Pew survey found that 84 percent of respondents said science had a positive impact on society. What's more, according to *Science and*

Engineering Indicators 2008, in annual surveys, more than 80 percent of Americans said they had "a lot" or "some" interest in new scientific discoveries. An overwhelming 87 percent expressed support for government funding of basic research, and a substantial 41 percent said the government spends too little on scientific research. On the other hand, your audiences also have reservations about science and technology, according to the NSF report. A majority agree that "scientific research these days doesn't pay enough attention to the moral values of society," and nearly half believe science causes life to change too rapidly.

Thus, although you enjoy the basic public support of science, your communications might not lead to blind acceptance of its benefits, says Cornell professor of science communication Bruce Lewenstein. "When scientists talk about public understanding of science, they almost always mean public appreciation of the benefits that science provides to society," he says. "Sometimes, though, greater public understanding could well lead to less support or more questions, and scientists must recognize that and accept it. If we take seriously the commitment to public understanding, what we are really taking seriously is our commitment to teaching people how to think and ask questions. And sometimes the answers to the questions won't be what you want them to be."

Appreciate That They Are Need-to-Knowers

Despite their interest, your lay audiences' knowledge of science is likely deficient. For example, a 2001 Harris Poll found that

- More than half of all American adults do not know that Earth goes around the Sun once a year.
- Nearly half do not have a sense of what percentage of Earth's surface is covered by water.
- Nearly half believe the earliest humans lived at the same time as dinosaurs.

What's more, much of the public believes in distinctly unscientific concepts. According to a 2005 Harris Poll, a significant minority believes in ghosts, UFOs, witches, astrology, and reincarnation.

However, concentrating on audiences' ignorance is not the most useful way to think about them, says Rogers. To enable the best communication, Rogers advocates focusing on audiences' experiential context. "People may not be able to define a molecule, but when they have a need to know about science, when it directly intersects with their lives, they have an amazing capacity to develop an understanding comparable to expert understanding in those areas," she says.

"For example, if you have a family member who is diagnosed with a significant illness you become expert on that illness, and your range of knowledge may go beyond that of many traditional experts."

So, rather than considering your audiences as simply willfully ignorant, recognize their need-to-know nature. Take responsibility to convey the relevance of your work for each audience. That relevance does not necessarily mean convincing them that your work will cure their cancer or make them money. Showing relevance can also mean convincing them that your basic study of quarks, DNA structure, or stress fractures is so interesting that they "need" to know about it simply because it is fun.

To be fair, you also suffer from ignorance—of your audience's knowledge of your field. Researchers consistently overestimate what the lay public understands about science. For example, Lehigh University's Sharon Friedman finds such overestimation when she briefs policy makers in Washington on media coverage of nanotechnology. "One of the reasons the policy makers keep inviting us back is to destroy the myth that the public really knows a lot about nanotech," she says. "People inside the Beltway and those in the nanotech research and policy areas all read the same stuff on the Web, they read *Small Times*, they read all the trade press, and they think that everybody out there in the world reads the same things they do."

Limit Their Conceptual Cargo

You may also vastly overestimate the ability of your lay audience to absorb information, especially if you are used to communicating intricate research concepts to colleagues. Lay audiences are just not willing to digest a multitude of complex concepts. Unlike a professional audience that (you hope) is tuned like a laser beam to your topic, lay audiences see your information—an article, a talk, a Web site, or a video—as just another piece of the constant cascade of information that inundates them every day.

"We are never listening or reading intensively," says University of Wisconsin professor of journalism and mass communication Sharon Dunwoody. "We are floating across the surface of text, even when we say we are playing close attention." Your audiences, she says, "will take only small bites out of what you have to communicate.... Think of audiences as individualistic need-to-know information-seekers sprinkled among people who don't care."

"The trick if you are communicating with these audiences is to figure out a way to get across one or two main points and do it well, because the rest of it will be ignored," says Dunwoody. "The problem is that scientists tend to have their peers in mind as they evaluate what information to include. They have a much more comprehensive notion of what belongs in stories, and if something is missing, they consider it a lethal flaw."

The context of communication also may limit their conceptual cargo space, says Lewenstein. In some cases, the context might dictate that you cannot even convey information, but only a positive attitude. For example, Lewenstein says, "If I take my kids to the science museum or the natural history museum, are they going to remember anything about the animals? No. But if all goes well, what they are going to remember is 'Dad spent time with me.'...So they will associate science museum with fun times with you."

So, a fair question is why bother to communicate to lay audiences if they will not follow you through all the fascinating (to you) intricacies of your research? What is the use of getting across a couple of major points? The answer is that even those couple of points can have enormous value for you and for your audience as well. If they stick in your audience's minds, they can become "memes"—the term coined by biologist Richard Dawkins for a "unit of cultural information" that can propagate from person to person, much like a gene. Such memorable memes can motivate an administrator to champion your research, a donor to write a check, or a student to join your laboratory. As a research communicator, I created many such memeish phrases—among them "artificial dog," "cosmic blowtorch," "anaconda receptor," and "shotgun synapse." You will have to wait until later in the book for their explanation, but for now just imagine how they might captivate readers of news stories and be passed along as shorthand descriptors for concepts.

Memes can also help propagate your research among fellow researchers, especially those in other fields. An engineer who visits a biologist's Web site might not be willing or even able to follow the details of a biochemical pathway. But that engineer will remember a vivid analogy about that pathway that could set him or her to thinking about an analytical approach to measuring that pathway.

Besides limiting the conceptual cargo you carry to audiences, you must also package it in standard ways, and if you understand them, you will be far more successful at explaining your research. Later chapters will cover these "packages" in detail, but for now here are a few examples to give you a sense of what they are:

- Television and radio reporters like quotes that are assertive, pithy, and memorable. The typical TV "sound bite" is only about nine seconds long, and the typical TV news segment a maximum of 90 seconds. So you need to prepare nine-second quotes that capture the essence of your findings. Otherwise, the editor will chop up your interview to get such a quote, and it might misrepresent your work.
- Legislators and their staff like "nuggets" of research news that they can use in speeches and with constituents. These nuggets are brief, compelling summaries of your discovery and how it will directly benefit voters.

- Donors also like concise explanations, but they also want a more personal connection with the work. For example, they like explanations of how you think your work will help people like themselves.
- Lay audiences prefer to learn about research, not just by reading text, but also through compelling photos, video, audio, graphics, and animations. So, to reach these audiences, you must also use these media to tell your research story.

Communicate within Their Values and Theories

In communicating with lay audiences, you must also take into account their pet theories, personal experiences, and core values, such as religious and political beliefs. Merely explaining your research concepts will not lead your audiences to "see the light" and appreciate your work's value, assert communications researchers Matthew Nisbet and Chris Mooney. In their 2007 article "Framing Science" in *Science*, they argue that people use their beliefs as "perceptual screens" and select the news outlets and Web sites that match those beliefs.

Thus, Nisbet and Mooney advocate that scientists should "frame" their information when describing research that involves a political or social issue. This framing should emphasize aspects of the issue that resonate with the audience's beliefs. This framing allows people to identify why the issue matters, who is responsible, and what action should be taken, they say.

Advocates are already using such frames, sometimes to the detriment of good science, warn Nisbet and Mooney. For example, skeptics of global warming have emphasized frames of "scientific uncertainty" or "unfair economic burden"—while those seeking to emphasize its consequences have used the frame of a "catastrophic Pandora's box." Similarly, antievolutionists have successfully emphasized frames of "scientific uncertainty" and "teach-the-controversy."

By contrast, scientists usually fail to resonate with the public when they emphasize only giving the facts. However, wrote Nisbet and Mooney, scientists discussing evolution would be more successful using frames such as "economic development," which focuses on negative repercussions for communities embroiled in evolution controversies, and "social progress," which emphasizes evolution as the basis for medical advance. Certainly, if your work sparks little controversy, you need not worry about such framing. However, when talking or writing about your research, keep in mind that audiences are screening your information through their values.

Lay audiences also hold pet theories about the way the world works that will color their perception of your communications. Some of these theories arise from our culture, and others from basic human psychology. Do not just ignore those theories, says Dunwoody. Otherwise, your efforts at persuasively explaining your research, eloquent as they may be, will fail. She cites as a classic example an epidemiologist trying to convince the public that a disease cluster is caused by chance: "It turns out, we rarely believe that, and part of the reason is that we humans tend to underestimate the likelihood that rare things will co-occur. Two seemingly rare things happen together, and we say it's not possible that it's chance; there has to be a reason."

Dunwoody counsels using "transformative explanations" to get around such a pet theory: "You first acknowledge the prevailing naive beliefs. You first tell the audience, 'Indeed when you see a cluster of rare things it makes absolutely no sense to suggest that could happen by chance. They seem so rare, that it just doesn't seem to make any sense.'" Only then, says Dunwoody, can you make the audience receptive to the truth.

Besides pet theories, the public also has a sharply different perception of risk than is warranted by statistical evidence. For example, the public vastly overestimates the risk of death from a sniper or terrorist attack or plane crash, compared to death from an auto accident or cancer—thanks to extensive media coverage of those dramatic and rare disasters.

Finally, your research explanations must take into account the "personal narratives" of audience members that color their perception of your information. For example, says Lewenstein, if you give a talk on breast cancer, that information is filtered through audience members' own history:

> They may have had breast cancer or have a relative who had it, and they may have a family narrative, which may be "Fight it with everything you've got because Aunt Tilly did and she lived another two years." Or "It doesn't matter what the hell you do everybody in our family has had breast cancer and died within six months. And so, why would I bother to do anything." That story, that knowledge, is as much a part of what they know as any particular biochemistry or clinical information you convey.

These personal narratives are cumulative, says Lewenstein, and this accumulation can have profound influence on people. He cites his own son Gabriel's interest in elephants. Gabriel's favorite stuffed animal as an infant was an elephant. And his youthful experiences included trips to zoos to see elephants, a visit with his father to an elephant preserve in Africa, the purchase of an elephant calendar, a scientific lecture on elephant communications, and scouring the Web for information on elephants. As a teen, he volunteered to work on an elephant communications research project. And as a collegian, he is considering majoring in environmental studies at Tufts University, whose mascot is...you may have guessed it...Jumbo the elephant. What triggered his lifelong interest in elephants? asks Lewenstein. "Was it the stuffed animal? Was it taking him to the lecture? Was it the trip to South Africa? Was it the calendar? Was it the availability of Wikipedia? The answer, of course, is yes."

Thus, to take this cumulative nature of personal narrative into account, your communication strategy should be to create a "herd of elephants"—not just one lecture, or a few news releases, or a quality Web site, but a spectrum of communications that give audiences many chances to experience your work in different ways.

Tell Them about the Other Side

Your credibility with audiences will be enhanced if you acknowledge those who disagree with you. Failing to discuss such disagreement—particularly when it represents a sharp division of opinion—could lead people to dismiss your position as naive or even disingenuous. Dunwoody suggests some graceful ways of making such acknowledgment: "I have seen people say, 'Having said this, there are some people who would disagree quite vociferously with me. Needless to say, I think they are wrong but let me share with you some of their points of view.'"

Many researchers are unwilling to discuss alternative points of view for fear that they may "infect" their audiences with incorrect theories, says Dunwoody. However, she says, "I suggest that people acknowledge even the flat-out nutty ideas, because they are out there. And the trick isn't to behave as though they don't exist—because that invites the reader to say 'I happen to know differently'—but to figure out a way to represent the idea that doesn't legitimize it."

She cites the example of explaining the controversy of evolution versus intelligent design: "A useful approach to addressing the controversy is to say 'You have heard of intelligent design. There are some people with scientific credentials who seem to be supportive of it. I will be happy to give you some of their names if you want. But I will say at the front end that this is a fringe point of view. There are more valid contested issues in evolution, and I would rather talk about those."

Learn to Speak Lay Language

Your audiences speak "lay language," not "research speak," and you must take that fact into account in your communications. Speaking this unfamiliar tongue might seem a bit foreign, even discomfiting. For one thing, you spend the vast majority of your time talking research speak with your colleagues. You share a specialized vocabulary and an understanding of your field's concepts. Since you are comfortable with that vocabulary and those concepts, you naturally tend to apply them to lay audiences—even when they might leave the audience utterly mystified.

So, overcome the temptation to lapse into technical jargon. It might be difficult, even when you try your best. Countless times, I have advised researchers before radio or television interviews to avoid technical terms, and they have nodded their heads in complete agreement and then proceeded to blithely bombard their interviewers with "enzymatic degradation," "stochastic analysis," and "hydroxyl ions."

You might believe that sprinkling a "retrovirus" here or a "quark" there in your lay-level explanations without defining them does not really compromise your communications. Surely, your audiences will forgive you an occasional unexplained term, and it will lend your explanations authority. But such jargon is a communications speed bump that interferes with their efforts to understand your work. It can also lead them to perceive you as arrogant and insensitive, turning off their interest.

You could even suffer serious professional penalties for this communication shutdown. For example, when explaining your work to your institution's president, you might believe it perfectly fine to use technical terms. But what if that president is not a scientist, but a political scientist or a Herman Melville scholar—as were two presidents I worked under? Foolishly peppering them with an unexplained "synaptic vesicle" or "Reynold's number" would likely lose a *very* important audience. What's more, such a president would leave that meeting perceiving you as a rather inept lay-level communicator, with possible impact on your career.

Despite these cautions, technical terms can have a place in your communications. If you explain them engagingly, they can be a spice that adds interest and authority to your story. Thus, "synaptic vesicle" could be explained as "a tiny ammunition pouch that holds the chemical bullets that one nerve cell launches to trigger impulses in another." "People may not understand scientific terms, but they expect to see some in research stories," says Dunwoody. "If you lop them out entirely, they will say 'This doesn't seem like a very scientific story.'" Thus, Dunwoody advises judiciously including only those technical terms really critical to the story, thereby distinguishing your tale of research from a tale of business, politics, or sports.

Besides avoiding technical jargon, be careful that your audiences do not misunderstand your use of words you believe to be lay language. As a prime example, scientists blithely use the scientific meaning of "theory" to characterize evolution, while their creationist opponents effectively use the word in its lay sense as a weapon to cast doubt on evolution. As Clive Thompson put it in an article in *Wired* magazine,

> In science the word theory means an explanation of how the world works that has stood up to repeated, rigorous testing. It's hardly a term of disparagement.
>
> But for most people, theory means a haphazard guess you've pulled out of your, uh, hat. It's an insult, really, a glib way to dismiss a point of view: "Ah, well, that's just your theory."

Thus, Thompson proposes that scientists alter their language when discussing evolution, referring to well-established science as "law," as in Newton's law of gravity, "because people intuitively understand that a law is a rule that holds true and must be obeyed.... Best of all, it performs a neat bit of linguistic jujitsu. If someone says, 'I don't believe in the theory of evolution,' they may sound fairly reasonable. But if someone announces, 'I don't believe in the law of evolution,' they sound insane. It's tantamount to saying, 'I don't believe in the law of gravity.'"

Thompson also warns that scientists shoot themselves in the semantic foot in using the word "believe":

> So when scientists talk about well-established bodies of knowledge—particularly in areas like evolution or relativity—they hedge their bets. They say they "believe" something to be true, as in, "We believe that the Jurassic period was characterized by humid tropical weather."
>
> This deliberately nuanced language gets horribly misunderstood and often twisted in public discourse. When the average person hears phrases like "scientists believe," they read it as, "Scientists can't really prove this stuff, but they take it on faith." ("That's just what you believe" is another nifty way to dismiss someone out of hand.)

Thus, in talking to the media and to lay audiences, consider replacing "believe" with phrases like "All our scientific evidence shows that..." or "We know that..."

Climate change communicator Susan Joy Hassol offered some excellent examples of the linguistic gulf between scientists and lay people, in an article in *Eos*:

> Scientists frequently use the word "enhance" to mean increase, but to lay people, enhance means to improve or make better, as in "enhance your appearance." So the "enhanced greenhouse effect" or "enhanced

> ozone depletion" sounds like a good thing. Try "intensify" or "increase" instead.
>
> "Aerosol" means small atmospheric particle to scientists but means "spray can" to lay people. "Positive" connotes good and "negative" connotes bad to nonscientists. So "positive trends" or "positive feedbacks" sound like good things. Instead of "positive trend," try "upward trend." Instead of "positive feedback," try "self-reinforcing cycle." "Radiation" is about X rays and Chernobyl for much of the public; try "energy" instead. "Fresh" means pure and clean, like fresh-smelling laundry; so instead of saying water will become "fresher," try "less salty."

In some cases, perfectly acceptable scientific terms might strike the public as odd, or even funny. For example, to avoid adolescent snickering during the 1986 *Voyager* flyby of Uranus, NASA scientists were allegedly instructed to pronounce it as "YOOR uh nus," rather than "your anus" in media interviews.

Learning lay language also will help your professional communications. Sensitizing yourself to how lay audiences perceive your words can only help you use and define them more effectively with all audiences, professional and lay.

Communicate *with* the Active Audience

Today, there is no longer an "audience" in the traditional passive sense of the word. Before the Internet, the "audience" only *received* information from the media. Now, with blogging, podcasting, online video, and Web sites, they actively *transmit and share* information.

The "active audience"—a term coined by BBC director general Mark Thompson—will instantly respond to your talks, news releases, articles, and books. They will post corrections and comments on news sites. They will blog and Twitter about a talk you are giving, *even as you give it.* They will create Web sites detailing arguments against your work if it is controversial. They may even digitally record and post a lecture or seminar, broadcasting worldwide an off-the-cuff remark you thought would remain restricted to the room.

The best way to cope with the active audience is to communicate *with* them rather than *to* them. A first step, says Carol Rogers, is to change your mental model of communications. "We have traditionally approached communications in a top-down or one-way communications mode, rather than as a conversation," she says. "The attitude is that we the research community are giving [the audience] information that we have, and they should understand it the way we want them to understand it." Rather, Rogers advocates a circular process, in which information flows both to and from the audience. That return route carries important information about how your audience is filtering the information through their perceptions.

Science communicator Rick Borchelt dubs the process "symmetric communications," and asserts that it produces better results than does the traditional one-way "asymmetric" approach. In fact, failing to use symmetric communications can mean more than lost opportunities to connect with audiences, warns Borchelt. It can cost an entire institution trust and credibility.

He cites a classic communications calamity in the failure of Brookhaven National Laboratory to tell the surrounding community that tritium was leaking into the groundwater underneath the facility in the 1980s: "Because they didn't think there was a great radiation risk, they decided not to tell the surrounding community, even though the contamination was creeping inexorably towards the surrounding community," says Borchelt. Invariably the leak came to light, and the Brookhaven management declared that, although they had known about the problem for years, they did not want to alarm the community, since the leak was harmless.

Harmless or not, the community was outraged at not being told. "This was in the West Hamptons where celebrities including Christie Brinkley and one of the Baldwin brothers have summer homes," says Borchelt. "So, they could readily get the ear of the Secretary of Energy and even involvement of the White House. DOE subsequently announced they were terminating the contract and firing the contractor for the lab."

Circular communications with lay audiences can also teach you about your own work, emphasizes Rogers. "Scientists' research is enriched when they interact with people who have an interest, because it may be one of these people who asks a question that leads them to see that 'the emperor has no clothes,'" she says. "You get so used to approaching something in a particular way, or going on a particular research path, because that is traditionally what your lab has done, that is where the funding is, etc. And then someone asks a question that could suggest another direction entirely. You never know when somebody might provide you with a spark of an idea that is so obvious in retrospect."

Pop Your Perceptual and Ego Bubbles

To really engage your audiences, you must pop a couple of rather comfortable bubbles that isolate you from your audiences. For one thing, you live in a perceptual bubble, says Lewenstein. He asserts that academics in particular

> just don't get it that most people are curious, but they are not curious the way university people are curious.
>
> Most people get up in the morning; they get the kids off to school; at the same time they get themselves ready for work. They have to drive forty-five minutes through heavy traffic to get to work. They go to a job which is a job; it is not something that is necessarily exciting or interesting. They get home at six or seven o'clock and they have to get food on the table; and then they have to get the kids to do their damned homework and practice the trumpet.
>
> And if they are really lucky they might have half an hour to sit in front of the TV and zone out without having anything else happening, before they go to sleep, and get not enough sleep, and get up and start doing it all over again. They are not interested in finding out about the current state of research on stem cells.

In your communications, pop this perceptual bubble; put yourself in the place of such overworked, overcommitted people and understand that you must compete with all their other distractions to get your information across. This perceptual bubble also affects your communications with professional audiences, says Lewenstein. "Even educated audience members are only educated in their field. If you are a biologist and you are not talking to a group of biology-oriented people, they are saying to themselves 'I never remember: cells, proteins, molecules, atoms; what order do they go in?'" Once you pop this perceptual bubble, you might appreciate more what your audiences really need from you.

You may also need to pop an ego bubble, says Duke research communicator Joanna Downer. "It's really hard to let go of that need to make sure that people know what you know. The way they do it is by speaking in terms that no one else can understand...such as using words like 'elucidate' that no one uses unless they are writing a scientific paper."

One way to pop these bubbles is to take every chance to talk to nonscientists about your work and listen to their responses. Practice explaining your work in an entertaining, engaging way at parties, family reunions, and lunches with nonscientist colleagues.

2

Plan Your Research Communication Strategy

Now that you better understand your audiences, you should also plan, at least informally, a research communications strategy. Even though such planning takes little time and effort, few researchers give their lay-level communications much thought.

"It is surprising to me that researchers will think nothing about jumping out of bed at three in the morning to come in to check on experiments; and they will do that day after day, year after year spending extraordinary amounts of time and effort," says Duke communication director David Jarmul. "Yet they won't invest even half an hour or so to think about this set of communications issues that actually may profoundly affect their research and their career."

Answer Some Key Questions

To help you develop a strategy, keep in mind these questions as you read this book:

- **Why do you want to explain your research?** To reach out to colleagues in other disciplines and to prospective students? To prepare yourself for a broader role as an administrator? To become a public spokesperson for your field or an advocate for a cause in which you believe? Answers to these questions will influence what kinds of communications you want to do and how you want to prepare yourself to do them. For example, some researchers may prefer to concentrate on communicating their work to their peers, rather than having it used in a more public way to further institutional ends. "They feel their contributions to their field are a very personal and individual thing," says Deborah Hill, communications director for Duke's Pratt School of Engineering. "And then, here is this whole organization that is trying to use that in ways that bring visibility to the institution and brings in money that may not go to the researcher. It can be uncomfortable to have your life's work and your persona leveraged in that way, and not everybody is going to want to play."

- **Are you a natural explainer?** If you are not, you need to decide whether to commit to training yourself to become a good explainer or just live with your shortcoming. If you are not a good explainer, take heart. Even Nobelists can be maladept at explaining their work. For example, journalist/author Keay Davidson calls Nobel Prize–winning astrophysicist George Smoot "hands down the worst explainer I have ever dealt with in my life." Davidson—who collaborated with Smoot on the 1993 book *Wrinkles in Time*—recalls Smoot's press conference after he received the prize.

> For the first two minutes he was lucid and excited and quotable, and then after that he went off into this vapor, and everyone in the audience was asking "What is he talking about?" I asked the first question because nobody else asked one, and I asked the second one because nobody else did. Later, at one point, he asked me "Keay, do you have any other questions?" I was the only one who knew what he was talking about, because it took me nine months to figure it out.

- **What vehicles do you prefer to use to explain your research?** There may be some vehicles that you simply prefer over others—whether they are news releases, magazine feature articles, op-eds, essays, videos, Web sites, and/or public lectures. This book will give you enough information to identify those that most resonate with your abilities and interests, as well as those that will yield the biggest payoff, given your goals.

- **How much time and effort are you willing to devote to explaining your research to lay audiences?** If you wish to restrict yourself only to exploring your research area to its fullest in the laboratory, you may not feel it necessary to spend as much time at lay communications. But if you see yourself ultimately becoming a leader in your field or institution, the time and effort spent explaining your research represents an excellent investment. Not only will it produce a trove of news releases, articles, videos, and so forth, that will serve your career, but you will also gain invaluable communications skills. And you will establish and refine the messages about your research, and even about science, technology, or medicine in general, that you think important to communicate to lay audiences.

- **Does your research field require explaining?** If you do basic research, with few applications of interest to lay audiences, you might safely restrict your communications to news releases aimed at professional media and other researchers. On the other hand, your work might naturally interest lay audiences, such as exploring the brain, or it might be a politically hot topic, such as the health effects of pollution. If so, you may find yourself in the public spotlight whether you want to be or not. In such a case, the most prudent course is to expend the effort to create communications that clearly explain your research and its broader implications.
- **Will you need protection from misinterpretation?** Will you find yourself in the public spotlight—for example, if your work involves invasive experiments on monkeys or has hot-potato political implications? If so, lay-level explanatory material will prove invaluable as a preemptive strike against miscommunication and misunderstanding of your work and your positions. Background materials are useful even if you do not have a newly published paper. The media may have questions about your work or about new findings by other researchers in your field. Well-crafted background materials will save you the time it takes to explain your work to multiple journalists or to other audiences. Also, standardized materials will help you avoid the kinds of errors in fact or interpretation that can prove frustrating and embarrassing.
- **Do you need to generate research support?** Maintaining support may mean persuading donors that your work is important enough to warrant their investment and persuading administrators to approve your research budget. It is much more comforting to walk into a donor meeting or budget review if the parties have read compelling articles or explored a professional-quality, informative Web site covering your work.
- **Do you need a "communications investment"?** If done wisely, your research communications not only will serve your immediate needs, but also will offer long-term payoffs. For example, at universities where I worked I made it a practice to write profiles on promising young researchers. These profiles—written before the researchers may have even published major papers—were printed in university magazines, posted on Web sites, and/or distributed to media as background. Those profiles found multiple uses—supporting award nominations, informing donors, and giving the university's administrators background on the hot new talents they had hired.

As indicated above, you need not address all these issues explicitly at this point. However, as you read this book keep them in mind. The answers will help you create a communications strategy that can advance your research and aid your professional goals.

Free Yourself from Suspicion and Risk

Fundamental to your strategy is freeing yourself from the attitude of "suspicion and risk"—as neurobiologist/communicator Chris Brodie puts it—toward communicating your work to a broad audience. You may view communication with suspicion, because of the imprecision of lay communication, says Brodie. "Unlike the graph or a piece of data where figuring out what it means is much less open to interpretation, you can construct words in ways that flip meanings around and obscure things." You might perceive communications as "risky," he says, because as a researcher you believe you are judged only on publication and funding, and that communicating beyond your peers poses a risk. And as for communicating with journalists? "Good God, imagine the mess that could be made if they get it wrong! Then people stop you in the halls and say [sarcastically] 'good interview' or say 'Did you really say that? What were you thinking?'"

However, even if you are a junior-level researcher and need to keep your head in the lab, neglecting to learn how to communicate at a lay level is short-sighted. You will immediately be faced with explaining your work to important lay audiences such as administrators and donors. And as you rise in the hierarchy, those communications will become even more important.

If you are already a senior-level researcher, by now you certainly clearly understand the value of lay-level communications to the success of your work and your field. More productive than "suspicion and risk" is a "protective-proactive" attitude toward your communications. For example, you should *protect* yourself by considering how a finding can be misconstrued or criticized when developing a news release or preparing for a media interview. But you should also relish the chance to *proactively* explain your work to these audiences, considering it an opportunity to define your research and its implications yourself.

Develop a Do-Tell Strategy

It is perfectly understandable that you may have given little thought to communicating your research findings beyond publishing scientific articles. However, just as a tree that falls in a deserted forest makes no "sound," research communicated only in the relatively unpopulated realm of the scientific journal makes less intellectual "sound" than it could.

So, develop a communication strategy of "do-tell"—that is, as you *do* your research, make it an integral activity to *tell* about it. Each time you achieve a research milestone, tell those who would be interested in knowing about it. Tell them when you win a grant, launch a new research project, install an important new instrument, form a new collaboration, establish a center, give a talk, or publish a paper. Your public information officer (PIO) can be a key ally in identifying such audiences and communications vehicles. And this book will help you manage the do-tell process yourself.

Develop a Strategy of Synergy

A "strategy of synergy" means making sure that you serve multiple audiences with the same communications. For example, most researchers consider a news release as aimed only at the media. However, as detailed in chapter 9, a release can also have a multitude of purposes and audiences, all of which benefit your work and your career.

Synergizing also means that if you are shooting a video news release, plan to shoot additional scenes for technical talks to illustrate an experiment or instrument. These videos might find wide use, enhancing your reputation in ways you did not anticipate. For example, if you are a biologist, your video might be featured on the *Journal of Visualized Experiments* Web site, which compiles Web videos of biological research techniques.

A good example of a strategy of synergy is Harvard Medical School's use of its research articles. Pieces written for Harvard's in-house publication *Focus* may also be adapted as news releases, donor-targeted research summaries, and material for the annual report. Also, the same article may be posted on multiple pages of the medical school Web site. For example, the school posts news releases both on the main news pages and on consumer-oriented "disease-based" research pages that highlight information on Harvard research into a particular disorder. These pages also draw content from the university's *Gazette* and the *Harvard Health Letter*. And, Harvard's communicators combine multimedia packages on Harvard research as "LabWorks." See the online resource section at *ExplainingResearch.com* for links to all these sites.

This strategy of synergy also has another meaning: the combined synergistic impact of your communications is much more than their simple sum. Your strategy should enable you to enhance that synergy.

Manage the Trust Portfolio

You are not just seeking to inform colleagues, educate students, and wow donors with your research communications. You are also helping manage your institution's "trust portfolio," the institution's reputation for integrity, credibility, and dependability, says science communicator Rick Borchelt, who has worked in government agencies, federal laboratories, and academe: "This portfolio is the trust relationships between our university, company, or agency and its external and internal stakeholders . . . just as the development office manages the donor portfolio; the CFO manages the financial portfolio; and the CIO manages the computing portfolio."

Your communication strategy should fit into your institution's or department's strategy of managing the trust portfolio to achieve its goals. And as a communications-savvy researcher, you should help your institution think strategically about how to allocate resources to support communications that meet the goals of the trust portfolio. For example, says Borchelt, if an institution decides to become a major player in stem cell research, scientists, administrators, and PIOs should plan to allocate their resources—to support Web sites, videos, feature articles, news releases, and so on—to communicate about stem cell research to enhance the institution's trust portfolio in that area.

Also (Ugh!) Market Yourself

Your research communication strategy also includes a marketing strategy. Perhaps you winced at the word "marketing," since it smacks of commercialism and salesmanship. Get over it. You conduct your professional career in a marketplace of knowledge and ideas, and marketing is precisely what you are doing when you are figuring out the best way to transfer your knowledge and ideas to those who need them.

Developing a marketing strategy involves analyzing your communication goals, identifying how to achieve those goals, and investing the necessary money and time. For example, in their marketing efforts, PIOs identify realistic media and other targets for news releases, Web sites, e-mail newsletters, and other materials. Marketing also involves measuring the results of that investment. For example, how many people visit your Web site? Who are they? What do online surveys reveal about what they liked and what they learned?

You will be a more productive researcher once you decide that you are not only a creator and communicator of new knowledge, but also a marketer of that knowledge.

You *Are* Media!

Your strategy should also recognize that you are now just as much a media outlet as any newspaper, magazine, or television station. You likely hold the misconception—also shared by some PIOs—that the media are the only important outlet for explaining your research to the public. This is a narrow 20th-century view. With the ubiquity of the Web and your access to media technologies and institutional resources, you can reach important audiences just as effectively as any traditional media outlet. For example, as indicated previously, your research news releases can appear in online news aggregators such as Google News and Yahoo! News, right along with media stories. Thus, your news releases, videos, and other communications need to have the same quality and credibility as do traditional media stories.

Apply Your Communication Strategy to Your Professional Publication

Your lay-level communications strategy can also inform your professional communications, although you might not believe it so. For example, you might believe your scientific papers need not be tailored to one scientific audience or another. However, professional audiences also comprise a range of researchers, and just as you tailor your lay-level communications for different audiences, so might you tailor your scientific papers.

Take the case of prominent Duke mechanical engineer Adrian Bejan, who developed "constructal theory," a seminal approach to describing shape and structure in nature. His theory can be used to understand systems as diverse as commercial cooling systems and lung vascularization. Despite the theory's importance, it initially failed to achieve the kind of broad recognition in the scientific community that it deserved. So, after initial news releases on Bejan's theoretical papers did not attract the deserved attention, Duke PIO Deborah Hill had a suggestion:

> He was quite frustrated, so I told him "Before I can really get attention for your work, I need you to publish scientific articles showing how your theory has broad applications." At that point, he didn't really understand the utility of what I was saying, but he went ahead and tried it. He published one article showing that the theory described how animals, from flying insects to fish, get around. And he published another showing how the theory predicts global circulation and climate. And he has since gotten tremendous media coverage and attention in the scientific community.

PART II

Effectively Reaching Your Peers

"I see by the current issue of 'Lab News,' Ridgeway, that you've been working for the last twenty years on the same problem I've been working on for the last twenty years."

3

Give Compelling Talks

Of course, communicating with your colleagues is your professional bread-and-butter. Failure to reach them might land you in the predicament of the fellows in the cartoon that begins this section—toiling away on the same problem only a few feet from one another.

Besides helping you communicate with your peers, this section of the book also aims at enhancing your communications with lay audiences. For example, technical seminars offer you a chance to practice the same audience communication skills—reading expressions, fielding questions, and extemporizing—that you can use in an interview with a reporter or chat with a donor. And even a jargon-stuffed technical paper gives you experience in writing a clear, concise sentence...or at least it should. And keep in mind that even your professional audiences might include "semi-laypersons"—scientists unfamiliar with your work. The same talk, news releases, feature articles, and Web sites serve to explain your research to both types of audiences. In crafting compelling talks, the subject of this chapter, you might believe that talking to your colleagues is very different from speaking to the local civic club. After all, when you present to your fellow researchers, you see yourself as Sherlock Holmes—using cold, hard logic to present your work. So adding the kind of P. T. Barnum–style showmanship to a technical talk that you might use with the civic club might not seem appropriate. However, a touch of carefully applied showmanship makes your professional talks livelier. And a lively talk makes your information more memorable and accessible, rendering your talk more effective. Also, even technical talks must persuade and inspire, and imbuing your talks with such qualities makes them—and thus yourself—more persuasive, credible, and authoritative.

Similarly, since professional audiences and the local civic club both comprise standard-issue humans, they will both respond to the energy and enthusiasm you inject into your talks. And you can hold the attention of both audiences by using the same clear elocution, dynamic phrasing and pauses, and memorable analogies, examples, and humor.

Besides similarities in technique, both types of talks follow the same strategy—to arouse and fulfill an audience's desire for information—says Tom Hollihan of the University of Southern California's Annenberg School of Communication. "You want to pique their interest, and then you want to satisfy that interest that you've piqued," he said in the video "Talking Science." "If you don't arouse them they never will get engaged and never connect and never listen. If you don't fulfill, they will walk away saying that wasn't a very satisfying talk."

You might mistakenly believe that you are already a brilliant speaker and do not really need to develop your speaking ability. After all, you are quite a hit with your colleagues when you give seminars. But in reality, those colleagues are probably all-too-forgiving of your shortcomings in both delivery and content. Such over-tolerance by professional audiences was vividly demonstrated in the classic "Dr. Fox educational seduction" experiment, reported in the *Journal of Medical Education.* In that experiment, researchers introduced a group of social scientists and educators to "Dr. Myron L. Fox," who gave them a talk on "Mathematical Game Theory as Applied to Physical Education." In questionnaires filled out after the talk, the audience declared the lecture clear and stimulating. But "Dr. Fox" was actually a professional actor, and his talk comprised a nonsensical mishmash of a few information nuggets from a magazine article, mixed with jokes, non-sequiturs, contradictory statements, and meaningless unrelated references. Nevertheless, since he wrapped the talk in impressive jargon, the audience judged his gibberish as valid information.

Free Yourself from Text

In giving a compelling talk, the most fundamental rule is to speak from points, rather than reading a text. While you routinely speak extemporaneously at your informal seminars, you might be tempted to prepare a text for more formal presentations. Resist that temptation. Reading from a text forces you to keep your head down, speaking into the paper instead of engaging with your audience. Reading also tends to reduce your voice to a monotonous, sleep-inducing drone and prevents you from moving around the stage—discussed later as important to maintaining audience interest. Instead of reading text, outline your points in whatever detail you need, and speak from that outline. As you practice, try to wean yourself even from that outline.

To make sure your audience gets all your points, you might provide them with your talking points or full text on your Web site or on paper. To develop a text from points, you could record your talk on a digital recorder, e-mail the file to a transcription service, and receive a text you can edit. You can also post the same audio online, as discussed later.

However, if you provide your audience a paper text, make it available *after* the talk, so the audience will not be distracted by trying to follow along with the text. Also, hold off on giving them the URL of a Web text. Given the ubiquity of WiFi, you might end up talking before a phalanx of open laptops.

An exception to the rule of not reading from a script is if your talk is very sensitive and precise wording is critical. If you must read your talk, write it to be spoken. Read drafts out loud, and try to make wording and sentences as natural as possible.

Organize Your Talk to Grab and Inform

Begin by organizing your points so they follow logically from one to another. You can organize your talk using the outline view of a word processor such as Microsoft Word, or specialized organizing/brainstorming software such as Inspiration. This progression should be logical enough that smooth transitions come naturally. Your outline will constitute a good start in developing your slides, discussed later.

Limit the number of points you cover according to the time available. For example, in a ten-minute talk, you can really only hope to cover three major issues, said Patricia Riley of the University of Southern California's Annenberg School for Communication, in the "Talking Science" video. "You can give the background on those issues, you can give some supporting data on those issues, you can explain why those issues are important," she said. "But if you try to cover seven issues in ten minutes, A, people won't remember it so you had better have very good handouts to give, and B, they will probably believe that what you have been doing didn't amount to very much because [in that time] you won't be able to give seven issues the impact that they deserve."

In creating your points, also keep in mind the three "rules of engagement" for getting your points across to audiences:

- Tell them what you are going to tell them.
- Tell them.
- Tell them what you told them.

Talks can have different structures, depending on your topic and objective. The how-to Web site Quamut offers a good list of structures:

- **Puzzle:** Construct an overall picture from topical puzzle pieces
- **Timeline:** Proceed through a chronological series of events
- **Spatial:** Discuss the topic by following the spatial organization of its parts
- **Questions:** Address five questions: Who? What? When? Where? How?
- **Order of importance:** Start with the least to most important details, or vice versa
- **Causal:** Explain cause and effect
- **Reduction of possibilities:** Eliminate alternatives to arrive at yours
- **Problem-solution:** Pose a problem and develop the solution
- **Thesis-antithesis:** Describe a thesis and then argue the antithesis
- **Logic:** Logically connect details to make a larger point
- **Yardstick:** List criteria you can use to evaluate your thesis or topic
- **Motivating:** Establish a need and then inspire the audience to act
- **Cicero's rules:** The classical Roman writer and orator Cicero outlined a classic framework:
    - Attention-getting introduction
    - Statement of the facts
    - Explanation of areas of disagreement or division
    - Support for your point of view
    - Elimination of the opposing arguments
    - Conclusion
- **Ron Hoff's structure**: The author of *Say It in Six: How to Say Exactly What You Mean in Six Minutes or Less*, offers this structure:
    - Introduce the issue
    - Give an overview
    - Present a solution with concrete evidence
    - Demonstrate the payoff of the solution
    - Interact with the audience

Once you have the basic structure of your talk, you can craft its components, in particular, an attention-grabbing opening, introduction, and summary. Your opening obviously will vary with the audience. For a technical talk, the most compelling opening is a concise statement of your research question. For lay audiences, you could open with a dramatic fact or statement, cartoon, striking image, joke, or anecdote. Such an opening tells the lay audience you are not going to inflict a dry technical talk on them. A later section covers using such elements effectively. For either type of talk, you can also pose an intriguing question that you will answer in your talk. Your talk's introduction should include a broader perspective on how your work or topic fits into an overall field and why it is significant. Do not be afraid to "show your ignorance" by discussing the limits of knowledge about your topic. Audiences, both lay and professional, love a mystery story. So, tell the mystery story of your work, the research challenges you face, and how you hope to solve the mystery.

Your talk's summary should send the audience away informed and interested in your topic. And, of course, they should perceive you as a credible, accessible authority on it. After the summary can come a slide with a simple URL of your Web site, where your talk and background on your work are available. A nifty way to reduce the length of URLs to make them easier to jot down is to use such URL-shortening services as SnipURL or TinyURL.

Make Your Talk Memorable

As mentioned above, even in technical talks do not hesitate to judiciously use humor, analogies, familiar examples, anecdotes, and quotes. People are people, whether professionals or laypersons, and they will respond to such features. Importantly, such features are more than just oratorical window dressing. They serve the central purpose of your talk—explaining your research—by holding the attention of your audience and making your presentation more memorable and your concepts more understandable. Of course, your humor should not sabotage your talk by making fun of your audience or being off-color or too self-deprecating.

Examples should be concise and on point. One good tip: to make your talk memorable, link your topic to current events. Has there been a recent dramatic discovery, disaster, or controversy in the news that you can relate to your topic?

Anecdotes about the surprises and challenges in your work also make your talk memorable. For example, cosmic ray physicist Dietrich Müller once told me a funny story about how his team developed a cosmic ray detector to be launched aboard the space shuttle. They were stymied about how to make the detector withstand the rigors of launch. As Müller sat frustrated in his office pondering what to do, he absentmindedly picked at the plastic fiber stuffing protruding from a hole in his winter jacket. In a Eureka! moment, he realized that mats of that stuffing would make a perfect cosmic ray detector. So the scientists tracked down the stuffing from the manufacturer, bought large quantities, and the high-tech detector was born. Besides making your talk more interesting and useful, such anecdotes can help portray science as a real, human endeavor with drama and serendipity.

Finally, a memorable talk ends with a bang, not a whimper. So, consider making the very last slide, rather than a list of credits or a photo of your research group, a visual that leaves the audience with a "wow"—such as a dramatic image of the creature you study, a beautiful galaxy, or a handsome picture of a bridge you discussed. Perhaps you might even use an artistic image, such as those discussed in the next section.

Use Engaging Visuals

Visuals enhance your presentation because people get most of their information through their eyes, not their ears. Thus, people more often say "I see what you're saying," than "I hear you." Studies have shown that 83 percent of learning occurs visually. These studies, described in a 1996 OSHA paper titled "Presenting Effective Presentations with Visual Aids," show that people retain about 10 percent of what they only hear in presentations, 35 percent of what they only see, but about 65 percent of what they *both* hear and see. Using visuals also makes your talk more persuasive. A Wharton School of Business study, cited in the 2004 white paper "The Power of Visual Communication," produced by Hewlett Packard, showed that 67 percent of audience members perceived as more effective a presentation that combined visual and verbal components. Both of these papers are available at *ExplainingResearch.com.*

So, make your presentations a feast for the eye as well as the ear. Follow the mantra of television news producers: "Say cow, see cow." If you talk about a cow, show images or video of a cow. And if you talk about a virus, an airplane wing, a quasar, or a bowel resection, show visuals.

Equations are an exception to this more-visuals-are-better rule. Communication experts recommend using as few equations as possible, even in a technical talk. Rather, concentrate on the concepts of your talk and leave the equations to a handout or your Web site. If you must use equations, keep them simple, label them clearly, and/or figure out a way to animate them using PowerPoint, so that the audience can see how they work.

Develop Slide Savvy

While slides are important to most presentations, they are not *the* presentation. Combine slides with your oral discourse, stagecraft, videos, demos, and pass-arounds to create a multisensory explanation of your work.

"A lot of scientists use slides as crutches rather than as means of telling the story," warns science communicator Ben Patrusky. In coaching scientists on their presentations, he often had to ruthlessly edit down their slides, especially to remove raw data. "People want to know trend lines, they want to see cartoons, and they want to see images. Slides should help advance the story for the audience and not serve as a crutch to click your thoughts in place."

Also, avoid "PowerPoint Phluff," which is what visual communication guru Edward Tufte calls ornamental design for the sake of design. While slides can usefully convey information, he says, they can also "replace serious analysis with chartjunk, overproduced layouts, cheerleader logotypes and branding, and corny clip art."

Do not list your points on extensive text slides. Rather, your points should be apparent in your slides' combination of text, images, animations, and graphics. Design your slides so that your audience will need to listen to you to understand fully what the slides portray. Make your slides so visual and so dependent on your talk that they would constitute an incomplete source if presented alone. One bad sign that your slides merely mimic your points is if you provide them as handouts, as if they represent the gist of your presentation. Handouts should be created separately as concise summaries.

Your slides should be visually dramatic enough to lure your audiences into listening to you and professional enough to enhance your credibility. So, try to avoid home-brew slides, especially for talks you will give many times. Have them professionally designed or use a commercial source such as ScienceSlides, which offers a broad collection of professionally designed slides. However, do not use canned slides, but adapt them for your presentation—for example, eliminating unneeded labels and diagrammatic detail. Also, explore the commercial sources for PowerPoint slide templates, such as PresentationPro. Unfortunately, many such commercial templates and graphics are PowerPoint Phluff. But while they would not be appropriate for a professional audience, they might usefully spice up presentations for such lay audiences as school or alumni groups. Lists of commercial sources of slides and other visuals are in the online resources at *ExplainingResearch.com*. Custom slides created using a professional graphic designer represent a good investment. The designer will usually come up with visual concepts you would not have thought of and can give you a template so you can add slides yourself.

As a general rule, organize your slides to include a title slide (with illustration!) and "mapping slides," which distinguish each section or point, and finally a conclusion slide. Also, slides should have enough information to warrant spending at least a minute for each slide, with more time for complex slides.

Become a "Power" PowerPointer

Learn PowerPoint thoroughly before using it for presentations. Take a PowerPoint course at your institution, go through online course and tutorials, or explore the software on your own. The simple step of accessing PowerPoint Help by pressing F1 will show you many useful shortcuts. Among them:

- F5 can start the show, Esc can stop it, and Shift F5 starts it from the current slide.
- To run a presentation, do not just open the file, which shows the slides in design view. Rather, right click on the file and choose Show. Or save the file as "type = PowerPoint Show" so you can double click on that file to open it as a slide show.
- If you are not using a remote, press the space bar to advance to the next slide.
- Press Control-Home to go to the beginning, and Control-End to go to end.
- Type a number and press Return to go to that slide.
- Use the commands to Hide and Show the pointer.
- Right-click to go to a slide by title and to show the pen, draw, erase, and hide the pointer again.
- Set up your show to start with a blank screen. Press B to make the screen black and W to make it white; pressing the same key again brings back your slide.

Check out the online resource section at *ExplainingResearch.com* for many good resources for better designing, using, and enhancing PowerPoint presentations. They include Microsoft's PowerPoint page, the PowerPoint FAQ, and PowerPoint Heaven. The resource section also lists software such as Articulate Presenter, and Web sites such as *SlideShare.net* and *SlideServe.com*, for sharing and distributing PowerPoint presentations.

Finally, invest in a wireless PowerPoint presentation remote control, which attaches easily to your laptop. It will unleash you from the podium, allowing you to roam the stage and still control your slides.

Become a Notable Keynoter

While PowerPoint is by far the most widely used software for scientific presentations, Apple iWork Keynote is also a highly capable slide production program for Mac users. For example, it features easy animation production and "Smart Build" image management. This latter feature enables multiple images to be nested into a single slide, with controlled transitions between images. Also, slide shows can be exported to QuickTime, so they can be played back on any computer, with controlled slide advance.

Other presentation programs to consider include OpenOffice.org Impress, Google Docs Presentations, and SlideRocket—the last two being online presentation-authoring systems that enable collaborative development. A particularly unique presentation program is Prezi, which can present visual information, such as a large diagram, as an overview on a large "canvas." You can zoom into specific areas of the overview to discuss details.

Avoid PowerPoint Pitfalls

Researchers encounter some common pitfalls when creating their slides. Avoiding them increases the likelihood that you will give an effective presentation. Among them:

- **Failing to give each slide a clear point that contributes to your story.** All too often the purpose of a slide is unclear. For example, researchers may present graphs without clearly labeled coordinates, making the reason for presenting the graphs unclear. Or, the labels are decipherable only to those involved in the research and baffling to novices. Chapter 4 covers in more detail how to create effective tables and graphs.
- **Using the default template of bulleted points.** Tufte argues that such bullets depict only a list of ideas and not relationships among the ideas. Rather, use diagrams, flow charts, and animations to reveal relationships and interactions. You can even animate PowerPoint diagrams, sequentially adding components and zooming in on specific components as you talk. And, as indicated above, there are many sources of software and tutorials for creating motion and animation. PowerPoint 2007 SmartArt graphics offer an easy way to create professional-looking, informational diagrams and flow charts, as well as to animate them. However, merely using animation to incrementally present unconnected information on a slide decreases learning, compared to having all the information shown at the same time, found a comparative study published in the *International Journal of Innovation and Learning.*
- **Using animation, 3D, or elaborate graphics for their own sake and not to advance your story.** For example, using fancy animated transitions between points distracts people and pulls them out of the flow of your presentation. But while you should keep technical presentations visually simple, a bit more slide show showmanship might better attract and hold lay-level audiences such as school groups.
- **Using all static slides.** Besides animated diagrams, video that advances your story is a terrific way to enliven presentations and communicate complex information. Learn to run movies and animations within PowerPoint.

- **Having unbalanced slide information content.** Avoid stuffing some slides with information, yet leaving others sparse.
- **Reading slides.** Your audience can probably read perfectly well. Also, reading your slides causes you to turn away from the audience and talk to the screen.
- **Showing text not meant to be read.** Presenters sometimes show a long block of text, for example, displaying a printed page to make a point about a book. Such an indigestible text chunk serves only to frustrate the audience and induce intellectual coma.
- **Using too many words.** Make your words count. A typical PowerPoint slide can hold only about 40 words, so tighten text as much as possible. As the saying goes, "Eschew obfuscation and eliminate redundant duplication." Use declarative statements and active verbs to give that text life. For example, change "Our work aims to..." to "We aim to..." or "Our aim:..."
- **Using an illegible font.** Use a sans serif font with a minimum size of 30 points. Readable sans serif fonts include Arial, Tahoma, Trebauchet, and Verdana. Although serif fonts are more readable in print, avoid them in slides. Serif fonts such as Courier, Georgia, Palatino, and Times New Roman do not render well on a computer display. Test your font by trying to read it from the maximum distance of the room in which you will present.
- **USING ALL CAPS.** Capitalized text is harder to read because it does not produce the recognizable "topographic" features of combined caps and lowercase text.
- **Using garish or noncontrasting colors.** Make your text dark on a light background such as an off-white or oatmeal. Take into account that many men are red-green color blind and may not be able to distinguish these colors if they are too close in shade.
- **Having poor information layout.** Design slides to be scanned upper left to lower right—the way people normally take in information. If your key point is at the bottom, people may miss it.
- **Skipping irrelevant slides.** Make your slides just fill the allotted time, so you do nor find yourself skipping through them. Each slide in your lineup should be necessary.
- **Pacing slides poorly.** Do not spend so little time on a slide that the audience cannot digest it, or so much that it becomes boring. Slides should be on the screen no less than ten seconds and usually no more than two minutes.
- **Overstuffing slides with data.** Researchers tend to show slides depicting a mass of data, believing that they may be excused for the numerical assault

if they just instruct the audience to "ignore everything but this line." They feel that inundating their audiences with numbers adds to their credibility, as if to say, "I really have a lot of data to back up my claims. I've really worked hard." Take the time to create slides that show only relevant data. You can still impress people with the size of your data set by posting comprehensive data on your Web site or as handouts, so people can analyze it at their pace.

- **Using too-complex tables.** Keep tables simple: no more than three or four vertical columns and six to eight horizontal ones. Studies have shown that people have difficulty grasping lists of more than about nine items.
- **Using a different laptop.** Presentations can sometimes present differently on different laptops. To be safe, run your presentation from the same laptop on which you developed the presentation.

Consider (Gasp!) Abandoning Slides

While slides are terrific for showing images and conveying complex information, at times you can communicate more effectively by scribbling on an old-fashioned black (or white) board or paper tablet. Projected slides, for all their usefulness, tend to direct attention away from you. Thus, they reduce the personal rapport that helps engage your audience. Drawing or writing text is an active process that pulls your audience into your arguments in a way that simply pointing to a slide cannot.

"PowerPoint itself may be one of the barriers we have to effective science communications," declares Carol Lynn Alpert of the Boston Museum of Science. "And yet…it is so embedded now in the culture of PowerPoint that a whole generation of scientists is being trained to speak to you through this medium of divided attention…instead of looking at you in the eyes and telling you a personal story about their research that they believe in," she told an audience at the 2008 AAAS meeting.

Your slides are not your presentation—a point Patrusky makes by recalling the case of a renowned scientist who failed to show up the night before a morning presentation at one of Patrusky's New Horizons in Science Briefings. "I called him at home," recalls Patrusky. "He answered the phone, and my heart dropped! He said 'I went to the airport, and when I put my briefcase down to sign my credit card for the ticket, somebody stole it. My slides were in it.' He went home because he didn't have his slides." Patrusky persuaded the scientist that he could explain his work without his precious slides. "I talked him onto a midnight plane; he arrived at 7:30 a.m. and was up talking at 8:30 with a piece of chalk."

Consider abandoning your slides for a blackboard or tablet when you are talking to relatively small audiences and when simple sketches and brief lists will tell your story.

Catch the Eye with Photos, Video, Cartoons, Animations

It is easy to give your presentation visual impact, drawing from the vast collections of free and commercial images and animations available. This ready resource means you need not settle for your own homemade images and graphics. And to develop images specific to your work, enlist the services of a graphic artist to render them.

For free images, check out the major scientific associations in your field as well as government agencies. Many times, they have ready-made background material that includes professional-quality graphics. Also, use Google's Advanced Image Search to find images about specific topics. The online resource section at *ExplainingResearch.com* lists many free and commercial sources for images, illustrations, animations, video, and science cartoons.

If you create your own animations and video, make them short. "A researcher might think short is five minutes, but unless you have a hugely compelling narrative story, five minutes in the world of video is an eternity," warns Tinsley Davis, executive director of the National Association of Science Writers, who managed presentations at the Boston Museum of Science. "Twenty seconds is probably enough to get your point across, especially if it is part of a talk."

Consider creating different versions of visuals for professional and lay presentations. Later chapters cover in more detail producing such videos, illustrations, and animations.

Give Memorable Demos and Pass-Arounds

Allowing your audience to handle a cool gadget or see a live demonstration could make your presentation unforgettable. For example, I vividly recall one speaker who used "throw-outs." The lecturer, an expert on integrated circuit design, wanted to emphasize the plummeting price of computer memory chips. To make his point, he pitched a fistful of the chips out into the audience, saying that even a few years ago such a chip-toss would have been incredibly expensive.

Make such pass-arounds relevant and understandable to your audience, however. A gadget that is arcane to your audience can do your talk more harm than good. Cornell's Bruce Lewenstein recalls a school parent's day, in which one parent, an engineer, gave a talk on his work that involved passing around a computer hard drive. The device mystified the eighth graders. "He only said something like 'This is what is inside your iPod,' and talked about the technical issues," says Lewenstein. "He didn't grab their attention or use a hook that would connect to them." Lewenstein, by contrast, showed pictures that explained how a toaster works, and "I was told later by my son and some of his friends that 'You were really interesting, but [the other boy's] dad was really boring.'"

Audiences also love and remember demos. For example, I still recall how my high school science teacher memorably demonstrated the concept of an explosive air-fuel mixture. He punched one hole near the bottom of a clean paint can and another in the lid. He filled the can with natural gas and lit the top hole, producing a small flame. As the flame flickered, he commenced to talk about what constituted an explosive mixture. Suddenly, the top exploded off the can with a loud bang. The air flowing into the bottom hole as the gas flowed out the top had ultimately produced an explosive mixture inside the can.

I also vividly remember renowned primatologist Jane Goodall demonstrating chimpanzee sounds by erupting with a bellowing "pant-hoot" in one of her lectures. The loud roar from the otherwise reserved Goodall woke up the audience, triggered a ripple of appreciative laughter, and brought her beloved chimps "into the room" in a dramatic way.

You might not have a pant-hoot to offer as a demonstration. And, you might not want to go as far as physics instructor David Willey, who famously walks on fire, dips his fingers in molten lead, and has a concrete block shattered on his chest while lying sandwiched between two beds of nails. But even more conventional demonstrations will enhance your presentation. For sources of demonstration ideas, see the online resource section at *ExplainingResearch.com.*

Demos need not be complicated to be memorable. For example, to create a memorable molecular model, rather than using a standard kit, use Styrofoam balls and "noodles," or perhaps even balloons. Visit your local craft shop to find unusual materials that could make for oddball, and thus memorable, demos. Certainly, do not be the kind of "demo-dunces" who miss golden opportunities to offer memorable demos. Ben Patrusky recalls some favorites:

> I had one speaker who I thought had a surefire story that everybody would enjoy—unisex clothing. But when he gave his talk, it was graph after graph, with not one picture or piece of clothing.
>
> And there was the speaker on bird communication. When I talked to him, he played bird calls and even did them himself. But in his afterdinner talk, he didn't play a single bird call, but showed nothing but sonograms.
>
> And another speaker on the psychology of eyewitness testimony, when I first interviewed him showed me a dazzling film of a crime being

committed, in which eyewitnesses completely misread the crime. It just hooked me, because I had also misperceived the scene. So, he gives his talk, and he doesn't show the film. I asked him about it, and he said "Oh, I forgot to bring it." I was stunned!

Practice Produces Powerful Persuasion

To ensure that your talk accomplishes what you want, practice, practice, practice—to which I would add *practice*! Practice strengthens the neural connections between your brain and your mouth, such that the words flow, rather than dribble haltingly from your tongue. Also, practice enables you to refine the content of your talk, timing it and perfecting the phrases that will most clearly and compellingly describe your work. As you practice, keep in mind the knowledge level of your audience, so you can adjust your content and language—for example, eliminating technical jargon.

Practice also helps you rid your speech of annoying uhs, ums, and y'knows. You might be stunned to find out how many you use—up to hundreds for even a short talk—unless you have assiduously banished them. And it helps relieve stage fright by giving you confidence that your talk will flow semi-automatically to make your desired points. If you experience stage fright, it might help to know that you are by no means alone. About three-quarters of Americans report suffering "glossophobia"—fear of public speaking—ranking it above even fear of death! A good way of overcoming stage fright and practice speaking skills is to join Toastmasters.

Practice the style of your talk as well as its content—for example, your energy and enthusiasm. Practicing before a full-length mirror will help you assess your posture and gestures. Your posture should be easy but erect and your gestures appropriate—for example, using your hands to depict how two enzymes fit together or how air flows over an airplane wing. Practice until you can move from point to point without consulting your notes and until all the phrases and concepts flow easily.

If your voice is not toughened by frequent speaking, you might want to whisper during the practice sessions, to avoid ending up croaking your presentation to your eventual audience. However, some run-throughs should be at full volume, so that you can judge whether your voice is loud enough.

Once you have a final presentation, assess it by recording yourself on video to review your performance. If you do not have a video camera, audio recording can at least tell you whether you sound as you want to. Also, test your presentation before friends or colleagues who represent your audience. To elicit useful feedback, do not just ask them "How was that?" or "Do you have any suggestions?" Rather, ask "What were the weakest points in my talk?" and "How can I make this talk better?" To test the effect of your final revisions, try another video or audio recording.

Besides recording your talk during practice, record your live talk, using video or a digital recorder. Not only can you critique yourself, but you might also capture turns of phrase that are better than those you would write, says journalist/author Keay Davidson: "I often discover that I can't express what I have to say very clearly until I have to talk about it in front of a live audience. Then it just comes out as clear as a bell, and I am so glad I taped it." When Davidson transcribed one talk, it provided the basis for the preface of a book he was writing. "Previously, I tried to write it down, I struggled and went back and forth and there were thousands of pages, and I never quite got it right," he recalls.

And as discussed later, consider posting the audio or video of the presentation itself on your Web site.

Use Good Stagecraft

A talk is also a stage performance, so here are pointers on delivery techniques, many from the Web sites SoYouWanna and Quamut:

- **Write your own introduction.** Your introducer will appreciate it, and it will emphasize the points you want made.
- **Try to scout out the room ahead of time.** This will enable you to suggest changes that will aid your talk: adjusting the stage, podium, seating arrangement, sound system, room lights, or room temperature. Make sure there are enough electrical outlets and that your computer connects properly with the projection system. Check whether your slides are legible on the screen provided. Ask for a bigger screen or rearrangement if it is too small. The screen should subtend a 20- to 30-degree viewing angle from the farthest point in the audience. Identify the technician who can correct problems.
- **Engage your audience beforehand.** If appropriate, chat informally with audience members before the talk. The discussion will give you a better connection with them, alleviate anxiety, and help you warm up.
- **Make eye contact.** During your talk, make serial eye contact with audience members around the room, to help maintain audience interest.
- **Be handy.** Do not jam your hands in your pocket, fold them in front, or clasp them behind. Let them hang naturally at your side, so you can

use them to gesture. Point with an open hand, rather than giving your audience a finger—it is much friendlier.

- **Nix filler words.** Do not let the uhs, ums, and y'knows that you eliminated in practice creep back into your talk.
- **Talk to the audience, not the screen.** When showing slides, turn to the screen only to point to something. Then turn back to the audience.
- **"Read" the audience.** Look for signs of confusion as you talk. If you see too many, stop and ask if people need further explanation.
- **Pause to refresh.** Between each main point, or after you have gone through a complex explanation or slide, pause for a moment to let the audience process the information. The silence will refresh their interest.
- **Use dynamic choreography.** Move around the stage while you talk, ideally using a wireless remote to control the slide show. Do not pace or fidget, but move purposefully to make a point. For example, when you discuss a slide, move toward it. Or, move to a different place when you make a new point, and/or step toward the audience to make an important or emotional point. Perhaps even purposefully walk in front of the projection occasionally. The flash of light reflecting off you will wake people up. One speaker even jumped on a table to dramatize a point about the "reality" of material objects like tables. (Of course, make sure you are agile enough not to take a pratfall!) Even consider walking among the audience, for example, during the question-and-answer period. Placing the audience between you and questioners physically immerses the audience in the question-and-answer process.
- **Change your pace and volume.** Audiences have an attention span of about 20 minutes. So, just as a good baseball pitcher changes his pitch to keep the batter off balance, reestablish your audience's attention every 20 minutes or so by talking faster or slower, louder or softer.
- **Use a laser pointer only if you cannot reach the screen.** The laser spot tends to jiggle annoyingly if your hand shakes at all. And if you zip the beam around and use it to circle things, the spot becomes especially annoying and distracting, like a pesky fly. Rather, touch the screen if you can reach it, or use your shadow to point to things. The sound of tapping the screen also tends to draw attention to your point.
- **Organize audience participation.** For example, ask for shows of hands on relevant issues, or even divide the audience up into clusters and ask each cluster to form an opinion or try to answer a question.
- **Have backup visuals.** Besides bringing your own laptop, have two backups: your slides on a separate flash drive, and overhead transparencies in case the digital projector goes down. In fact, given that

digital projectors are cheap and small, consider buying your own if you are going to give frequent talks.
- **Yield to food.** You cannot give a good talk while people are going through a buffet line or eating. Schedule the talk after the meal and dessert.

Manage Questions

Note that the title of this section is not "answer questions" but "manage questions." While you should certainly *answer* questions, your aim is also to *manage* the question-and-answer period to explain your research authoritatively and to maintain your credibility. Here are tips on managing questions—many adapted from the booklet "Giving Talks" by the Burroughs Wellcome Fund—that will enable you to achieve those goals:

- **Anticipate the worst questions.** Think of the thorniest, most embarrassing questions you could get. Either incorporate the information into your talk or prepare answers for those questions. It is better to admit, "Indeed, I did drop an important fossil and smash it, so my data are incomplete," rather than have someone bring up your clumsiness. Be prepared to address questions from your most ardent competitor.
- **Practice accessing your slides.** Practice accessing specific slides to answer a question, so you do not spend time clumsily clicking through them.
- **Tolerate interruptions.** If you invited questions during your talk, be willing to stop and answer them. However, if someone interrupts you in the middle of a point, feel free to politely ask them to wait until you are finished with the point. Do not just ignore raised hands.
- **Repeat the question.** This guarantees that everybody in the room heard the question and that you understood it correctly. It also gives your brain time to percolate a good answer.
- **Understand the question.** If you do not understand a question, ask the questioner to repeat or amplify. If you still do not completely understand the question, say "If I understand what you're asking, here's an answer." The questioner can correct you, if you are mistaken in your understanding.
- **Make up your own questions.** To fill embarrassing silences and take the initiative, formulate a few questions you want to answer yourself. For example, if you noticed puzzled looks when making a point during your talk, expand on that point, saying "One thing I might not have been clear on..." This shows the audience you are humble about your communications and really care about their understanding.

- **Admit when you do not know the answer.** But do not just leave it at "Gee, heck if I know." Explain why you do not know and what can be done to find out. Also, be explicit about whether you personally do not know something, or your field does not know.
- **Give yourself a verbal escape route.** If a question requires a too-lengthy or complex answer, say "Here's a short answer, but we can talk about that in more detail together," or ask "Can I get back to you on that?"
- **Be prepared to answer questions you already covered in your talk.** Even if you have just made the most crystal-clear talk ever, some audience members will have just zoned out. Explain the point a bit differently, to give audiences a different take on it.
- **When attacked, be gracious.** Do not be defensive or hostile. Being polite and showing your patience and your tact will leave a positive impression with the audience. Take the attitude that argument is part of the give-and-take of research and is, in fact, an enjoyable, stimulating part.
- **Be kind to the ignorant.** If you are giving an astronomy talk about the accretion disks around black holes and a questioner asks "What's a star?" be kind—assume the person is not stupid, but lacks background, is inquisitive, and is brave enough to ask the question. Be very generous in explaining what a star is. In fact, even such basic information might contain insights that the astronomers in your audience have not thought about.
- **Tolerate the crank.** If you get a just-plain-goofy question, smile and answer as best you can, remaining respectful to the questioner.

Give Your Lay Audiences What They Want

Preparing a lay-level talk requires more audience homework than does a technical talk. While your technical audience is sophisticated and relatively homogeneous, your lay audiences can differ considerably in age, education, religion, and so on. Such factors will affect how you present your topic. For example, you would explain evolution very differently to religious fundamentalists than to an alumni group.

Also, audience expectations and the occasion of the talk will affect your talk. Are you talking to a class of motivated, retirement-age people who will enjoy a challenging intellectual engagement? Is it a group of foundation administrators who need to be persuaded to support your work? Or, is it a student group that wants to be wowed by fascinating facts and visuals in an afterdinner speech? Besides such strategic factors, here are some useful tactics for lay-level talks:

- **Establish your credibility.** Make sure your introduction by the host lays the groundwork for your credibility. And begin your talk with authoritative sources and explanations to firmly establish your authority.
- **Adapt your style and tone to your audience.** You can be more colloquial and informal for a lay audience than for a professional one. As to tone, when giving a speech to schoolchildren, for example, smile and appear more friendly and warm.
- **Do not talk only to a segment of your audience.** For example, if you have colleagues in the audience, you might be tempted to skew your talk to them, like the unnerved physicist in one seminar run by Ben Patrusky. "I worked with him carefully to simplify, simplify, simplify," recalls Patrusky. "And he looked out at the audience at [Nobel Prize–winning physicist] Shelly Glashow sitting there, and it turned into a postgraduate doctoral thesis for one. I was amazed. He was just so intimidated by the presence of Glashow, he just went berserk."
- **Do not simply repurpose your technical seminar.** Besides eliminating technical slides and jargon, do not expect to discuss your latest findings in much detail, says Tinsley Davis. "The topics that were most well received by our [Boston Museum of Science] audiences were those that, unfortunately, the scientists had to rethink how they told the story of their research. And they would be disappointed because they realized that fifteen minutes of their allotted twenty minutes had to be background, in order to reach the public."
- **Give lots of background.** Says Davis, "People are more likely to remember the talk, because if you give them a context to understand your field, then not only are they going to understand your latest work during that last five minutes; the next time they read something about you or your field in the newspaper, they are going to remember that background." Refer people who do want details on your latest work to your Web site.
- **Emphasize the "so-what" of your topic.** Tell explicitly why your work is important and interesting. This so-what may be different for different audiences—donors, alumni, students, and so forth.
- **"Lie" accurately.** As a researcher, you might feel that skipping details of your work is tantamount to lying. However, omitting details that can

safely be glossed over is sometimes necessary to achieve a larger goal—clearly explaining the basic concepts of your work. Remember from the discussion on audiences (see chapter 1) that you need to limit the conceptual cargo you load onto them. And you can always refer audiences to your Web site for such details.

- **Use concrete language and images.** What do elements of your work look like, sound like, taste like, or smell like? Speaking to scientists, you might not describe, say, the smell of a poisonous mushroom or the hiss of a cobra, but such sensory detail will captivate a lay audience.
- **Emphasize emotion and other human traits.** You are not giving a professional presentation, in which you are expected to purge every shred of humanity in favor of cold, hard logic and data.
- **Tell stories.** As journalism lecturer Carol Rogers says, the art of storytelling is one that unfortunately has not been well cultivated in science communication. "People are storytellers," she says. "That is traditionally how we have conveyed information. But somehow in the whole professionalization of science communication we have forgotten that. We left out the storytelling element, not understanding that people who aren't expert in science really relate to stories, and that the most successful communicators are those who tell stories."
- **Be colloquial.** Use slang, even jokes, and puns. It will connect you to your audience, and this connection will help get your points across. The online resource section lists sources for science and engineering jokes, and you likely have some favorites of your own.

Finally, see your lay-level talks as a great chance to practice talking to such important audiences as the media and even your colleagues. For example, Ohio State research communicator Earle Holland recalls the happy spinoff of glaciologist Lonnie Thompson's first major public talk on his research in Antarctica: "He was scared to death. He had only given a half dozen media interviews at that point and hadn't done any talking outside of teaching," recalled Holland.

> He gave a great public talk to about three or four hundred people. About two days later, I interviewed him for a research story, and his ability to explain his work was about two orders of magnitude greater than before. There was more content, and he was more open and willing to prognosticate. And a few days later, when he gave a brown-bag lunch talk to other faculty, he was also far more interesting than before. He was much more willing to be open about what was going on in his mind and how he saw his science developing.

Synergize Your Effort

Per your strategy of synergy, post recordings of your talks on your Web site as a streaming video or an audio coordinated with your slides. You can also link to data, background documents, and other sites. Such sharing will enable people to further explore your work and also draw traffic to your Web site.

High-end software for creating such online lectures include Echo360 and Sonic Foundry Mediasite. Also, Adobe Ovations and Articulate Presenter enable production of narrated Flash-based lectures from PowerPoint slides.

And as indicated earlier, you can share your presentation on the Web by posting it to *SlideShare.net* or *SlideServe.com*. These free services enable you to post slide shows, as well as create "slidecasts" in which your slides are synchronized with audio of your talk. Those sites also enable you to embed your presentation in your own Web site.

4

Develop Informative Visuals

Far too many researchers rely on amateurish charts, graphs, diagrams, and animations to explain their research, when well-done visuals would communicate more clearly and give their work a professional appearance. It is perfectly understandable that you might neglect good visuals. After all, they usually need frequent revisions to reflect new findings. And you might also worry that your colleagues would criticize your efforts as hyping your data by making it "glossy." Get over these qualms. At least consider learning enough about PowerPoint and graphics packages to make your homemade visuals high quality. Better still, pay for a professional artist to create and maintain quality visuals depicting your work.

Invest in Evergreen Visuals for Multiple Uses

Investing in professional-quality visuals is likely well worth it, given the payback. For one thing, some might be evergreen enough to use for quite a while before updating. Also, your visuals will reach multiple audiences—colleagues, students and prospective students, administrators, donors, the media, and so forth. What's more, given the reach of the Web, your quality visuals posted online might become standard references, giving broad exposure for you and your work.

For example, both the Why Files and the NSF routinely feature illustrations from researchers on their Web sites.

Of course, an image most directly benefits your research if it is striking enough to make the cover of a journal. One excellent example of such a photo that paid off was an image for a paper in *Cell* by Howard Hughes Medical Institute (HHMI) investigator Charles Zuker and his colleagues. Their paper covered the functional identification of the mammalian sweet-taste receptor. To garner the cover, Zuker hired a professional photographer to shoot mice nibbling at a luscious-looking chocolate pastry. The cover was striking, and it called attention to the paper.

Regarding such investments, HHMI investigator Thomas Steitz said in an article in the *HHMI Bulletin*, "I don't think anybody will get a job or be promoted because he or she had a cover; it's what's behind the cover that will get [a person] promoted. Still, it increases the impact factor—and that's very important." What's more, HHMI's Web news site, like other such sites, frequently uses such images to call attention to a news release on a paper. "Depending on the copyright, and if we can get permission from the journal, we will use such images," says HHMI public information officer (PIO) Jim Keeley. "And even if an image is not selected for a cover, and the investigator owns the copyright, we will still use it," he says.

Once you begin to focus on creating visuals, you will likely find unexpected benefits in understanding how to communicate your work. Former *American Scientist* editor Rosalind Reid recalls such an Aha! moment that occurred to a scientist in a workshop she gave on research illustration. "We had an exercise in which the participants were asked to just scribble some drawings to depict their research. And one guy really got into it, making a bunch of drawings. At the end of the workshop, I was hanging them up to show, but he wanted to keep his. I asked why and he said 'I am taking this back to my dean, so I can finally get him to understand what I do!' He put it under his arm and marched off."

To seek such quality visuals, explore the free and commercial sources listed in the *ExplainingResearch.com* online resources for chapter 3 on giving a talk. You might also browse these sources for ideas to adapt for your own visuals. For particularly inspiring examples of visuals, see the Best in Show section of the site.

Create Graphs and Tables That Reveal the Data

Of course, graphs and tables are the basic visual stock-in-trade of research. Fortunately, observing only a few basic guidelines will enable you to produce quality graphs and tables that communicate your research clearly.

Three excellent books on visual communication of research are Edward Tufte's *Beautiful Evidence* and *The Visual Display of Quantitative Information*, and Stephen Few's *Show Me the Numbers: Designing Tables and Graphs to Enlighten*. Many of the guidelines in this chapter are gleaned from those books. A good guide to statistical and graphical tools for depicting scientific data is *Visualizing Data* by William S. Cleveland.

Maximize "Data-Ink"

Effective graphs and tables maximize "data-ink," wrote Tufte in *The Visual Display of Quantitative Information*. He defines data-ink as "the non-erasable core of a graphic, the non-redundant ink." Tufte wrote that "every bit of ink on a graphic requires a reason. And nearly always that reason should be that ink presents new information." Thus, he advocates that designers of graphs and tables "erase non-data-ink, within reason." As you read through this chapter, you will see how this principle can guide design.

One important exception to Tufte's maximize-data-ink principle: To engage a numerophobic lay audience, consider ornamenting tables and graphs, at least modestly. For example, why not use cartoon faces for points on a graph, or candy bars to depict bars on a bar chart? Although Tufte would decry such additions as increasing non-data-ink, they could help engage a lay audience to understand the principle behind the data. And their humor helps retain the audience's attention.

Remember the Memory

Design your tables and graphs to "remember the memory." As Few points out, the brain's temporary short-term memory—which your audience relies on during your talks—has limited capacity. It can hold only between three and nine items. So, limit the components of your tables and graphs, and also limit the points you want them to make. For example, a bar chart with a dozen bars will be more effective if you divide it into separate charts with fewer bars—if such a division is logically reasonable.

When a Table, When a Graph?

Use a table when your audience needs to see the numbers, but use a graph when they need to see the relationships among those numbers—the *shape* of the data. As Few wrote in *Show Me the Numbers*: "We can't take a bunch of numbers from a table and chunk them together meaningfully for storage in short-term memory;

we can, however, discern in a graph or image of a single, meaningful pattern that is made up of thousands of values."

The brain encodes graphs and tables differently, Few points out. The brain encodes graphs visually and tables verbally. This implies that you should use tables sparingly, since your audience members primarily use their visual sense in grasping your points.

Make Tables and Graphs Guide the Eye

Design your tables and graphs to "guide the eye." That is, their layout should cause a viewer's eye to naturally gravitate to your visual's main point or follow the sequence you are trying to explain.

While it's perfectly fine in a lab meeting to present a table or graph with undifferentiated hordes of numbers, you risk losing your audience if you do so in a seminar, symposium, or public talk. So, tailor your tables and graphs to the presentation. For example, it only takes a little time to produce a custom table with the relevant column shaded differently and/or irrelevant columns eliminated.

Some other general principles for designing tables and graphs to be easily understood:

- **Group similar data, such as columns in a table.** Use white space to separate them or enclose them in lines.
- **Avoid using heavy grid lines in a table.** They break up the data in tables and confuse the eye in graphs. Remember: "erase non-data-ink." Use thin dividing lines if necessary to enable lookup of numbers.
- **Do not use grid lines in graphs.** It is the shape of the data that is important.
- **Make axes, ticks, and labels less prominent than data.**
- **Use line width, size, or color intensity to highlight the most important data.**
- **Position data in tables or graphs to emphasize their importance.** For example, data positioned at the top left will be given more weight than that at the lower right.
- **In positioning data, think about what you want the viewer to do with them.** For example, place numbers to be compared next to one another and highlight them.
- **In graphs, try to label lines directly.** If you must use a legend, place it as close to the line as possible, so the viewer's eye will not have to jump too far.
- **Do not use pie charts.** Their form makes it difficult for viewers to intuitively compare the areas representing different quantities. Rather, use a bar chart.

- **Do not use 3D in graphs.** It merely adds non-data-ink.
- **For complex graphs, distinguish data points by using shape, size, or color.** Distinguish lines by using color or thickness.
- **For line graphs, avoid using varied line styles.** For example, do not use long dash, short dash, and dot-dash in the same graph, unless the graphs will have to appear in print in black and white.
- **For bar charts, use horizontal bars when displaying a ranking relationship in descending order or when labels are too long to fit side by side. Use vertical bars when data are to be compared.**
- **Start the axes of graphs at zero to avoid "lying to the eye."** Or, clearly label the fact that your graph does not start at zero.
- **Similarly, do not distort graphed data by choosing axes, for example, to overemphasize changes in the y-axis direction.**

Be Carefully Colorful

Unless your visuals must be in black and white, color can be invaluable in making them effectively convey your point. Some general principles:

- **Establish a "color code" for groups of visuals, such that a given color stands for the same kind of quality.** For example, your key information might be in red throughout a presentation.
- **Use colors that differ significantly in brightness as well as hue, so that they will photocopy distinctively.**
- **Rather than using a random mix of colors, use a natural progression of shades.**
- **Use subtle shades to guide the eye.** For example, shade a column of numbers you want to highlight.
- **Use more saturated attention-getting colors to emphasize main points.**
- **Make sure colors contrast well with backgrounds or other colors, for example, colors in neighboring bars on a graph.**
- **Design the area of color to be large or vivid enough to be perceived.** For example, the colors of narrow graph lines should be vivid enough so they will not be mistaken for black.
- **Since many men have a red-green color-blindness, use either red or green but not both.** Or, vary their intensities enough that even color-blind people will see them as different shades.
- **Consider the cultural background of audiences.** For example, red signifies danger in Western cultures but good fortune in China.

- **Use color rather than fill patterns on bar graphs.** Cross-hatching, striping, and other such patterns create a dizzying moiré pattern when next to one another.

Make Text the Medium of Your Message

Some general principles for effective use of text in your visuals:

- **Edit text in your tables and graphs as tightly as possible, just as with PowerPoint slides.**
- **Use full text to label the elements of graphs, for example, their axes.** Do not use text abbreviations or mathematical expressions as axis labels.
- **Use figure captions to give the conclusion to be drawn from that figure, not just to label the figure.**
- **Use readable fonts.** Keep in mind that, while serif fonts are more readable in print, sans serif fonts are more readable on a computer display, since the display cannot render the serifs well. Among the candidates:
    - Courier is a legible serif font.
    - Georgia is a legible serif font.
    - Palatino is a legible serif font.
    - Times New Roman is a legible serif font.
    - Arial is a legible sans-serif font.
    - Tahoma is a legible sans-serif font.
    - *Trebauchet is a legible sans-serif font.*
    - Verdana is a legible sans-serif font.
    - *Script fonts such as P22Corinthia are not legible.*

And do not mix your fonts. *It drives* audiences *nuts!*

Finally, do not use vertical text. It gives viewers a pain in the neck.

Explore Graphing Software

Adobe Illustrator is a widely used software package for producing graphs, and Adobe provides comprehensive online tutorials. Even if you plan to use a graphic artist to create your graphs and charts, taking the tutorials to familiarize yourself with the package is well worth it. Be aware that Excel—widely used for graphing—has endemic flaws, offering too many colors, distracting backgrounds, bad text formatting, and 3D bar graphs.

Illustrate and Animate Your Work

You can create good basic illustrations and animations yourself, using such resources as the Microsoft PowerPoint site, PowerPoint Heaven, and add-on software such as Ovation. However, consider a professional artist for elaborate illustrations and sophisticated Flash animations. The investment is worth it, because the visuals will convey your work far more effectively in your talks, Web site, and publications.

However, even if you work with an artist, familiarize yourself with graphics and animation software, so you will understand the process and possibilities. For example, Adobe offers good tutorials on Flash animation and the use of Photoshop to animate scientific images.

When seeking an illustrator or animator, first check your own institution. Your PIO may help identify in-house artists and also may work with you and the artist to create visuals for both technical and lay audiences. Your PIO can be valuable in offering a third set of eyes and ears and a practiced expertise at explaining research that neither you nor the artist likely has. And, a PIO also can serve as a translator between scientist and animator, relieving the researcher of much frustration, says Vanderbilt PIO David Salisbury. "Often, it is not enough to put a graphic artist together with a scientist," he says. "They can have trouble communicating. I try to act as a middle man between the two and manage the process." Your homegrown graphics and animations might provide the basis for lay-level illustrations, says Salisbury, who has adapted many researchers' visuals for the university's Web magazine *Exploration*. "Often, we only need to replace the enigmatic scientific labels with English and come up with captioning that explains what's going on," he says.

Some general guidelines for working with an artist and/or animator:

- **Check out free and commercial artwork.** For a list of what is available and to get ideas for your own visuals, see the list of sources in the references for chapter 3 on the *ExplainingResearch.com* Web site.
- **Before you approach an artist, brainstorm with your PIO and colleagues about what the illustration or animation should show.** If the visual will be used in a lay-level presentation, ask for ideas from people who represent your potential audience.

- **Give the artist something to start with.** This so-called art scrap can be your own PowerPoint slides, hand-drawn sketches, or sample visuals from outside sources.
- **Have a two-way discussion with the artist.** Tell the artist what concepts you want to convey but also listen to his or her ideas. Appreciate that those ideas come from a sophisticated visual communicator.
- **Understand the costs.** This initial discussion should enable the artist to come up with an estimate.
- **Understand how frequently the visual will need to be updated and the costs involved.**
- **Expect to iterate.** Even with the most productive initial discussions, an illustration or animation will invariably go through multiple iterations.
- **Do not be timid about asking for changes.** Software for drawing and animation makes such changes quite easy.
- **Audience-test the visual before finalizing it.** Ask colleagues, family, and friends to review the illustration or animation, and use their feedback to refine it. Ask what they understand from looking at the visual and what confuses them.

Despite the effort and cost, high-quality tables, graphs, illustrations, and animations will offer both short- and long-term payoffs. Not only will they enhance your research communication, but what you learn about "visual language" in developing them will serve you throughout your career.

5

Create Effective Poster Presentations

Poster presentations, like professional talks, can be more broadly useful than just to display your findings at conferences. You have no doubt seen posters mounted in laboratory hallways and at receptions and lay-level meetings. What's more, you can share your posters on the Web, using such services as *SciVee.tv* "postercasts."

Thus, produce your posters not just to inform colleagues at a conference, but to reach people beyond your field.

Create an Accessible Design

To create a good poster, you need a word processing program, plus graphics packages such as PowerPoint, Macromedia Freehand, Adobe Illustrator, Adobe Photoshop, or Adobe PageMaker. If you have not created a poster with such programs, ask someone whose poster you liked to share their layout, and adapt it. However, also consider having your posters professionally designed, especially those you will use repeatedly.

In either case, start with a rough draft and ask friends and colleagues to critique it by attaching sticky notes. If you convert your draft to a jpeg and upload it to the photo-sharing site *Flickr.com*, people can add electronic sticky notes to it, suggests Swarthmore College professor Colin Purrington. A link to Purrington's poster design Web site and other sources for poster design advice and templates are included in the online references for this chapter at *ExplainingResearch.com*. Here are some design guidelines from these sources:

- **Make your layout regular, lining up the elements.** In organizing the layout, remember that the eye travels from top to bottom, left to right. So use a column layout, with the sections arrayed top to bottom, then left to right. Making viewers read laterally across the poster before returning to the left will drive them to the next poster.
- **Array your most important information at the top.** People in front of your poster will block information along the bottom. This "grabber" material at the top should hook audiences to stop and engage your poster and you.
- **Use dark letters on a white or neutral background.**
- **Use only use a few colors and use them consistently to denote a specific element or category.**
- **Keep column width narrow and paragraphs short; no more than 50 to 75 words.**
- **Do not jam the poster with text and graphics.** Rather, use white space and graphics to guide the reader and give the eye "breathing space."
- **Use no more than two or three items per list.**
- **Make captions or paragraphs no longer than the column width.**
- **Make text size readable up to six feet.** Use a sans serif font, upper- and lowercase, for easier reading. Use a hierarchy of text sizes and bolding and graphic size to signify importance. Single space text.
- **Make graphs clean and uncluttered.** Use as few numbers as needed to give viewers necessary information. Your Web URL can guide them to complete data.

Tighten Text, Create Grabber Graphics

Because text is very limited on a poster, make every word count. For example, the headline should be a clear, tightly worded statement of your conclusion, not what you did. Beneath the headline, list your identification and contact information and Web site URL. Acknowledge your funding support at the bottom of the poster, since it is not as important to your audience.

"Microedit" text ruthlessly to tighten wording. For example, tighten:

- "administration of" to "administering"
- "transplantation of" to "transplanting"
- "schematic illustration of" to "schematic of"

Use short sentences and avoid passive voice. Researchers seem to have a genetic defect that renders them incapable of either. Prove them wrong! Chapter 6 contains a collection of microedits and writing tips.

While you might organize your poster into Abstract, Introduction, Results, Methods, and Conclusions, do not use these headers—they are boring. Rather, make each subhead a statement that summarizes the section.

The introduction should state why your work is important and how it fits into the existing research paradigm. Aim this perspective at someone who knows little about your field. Do not focus excessively on your experimental methods. People can obtain them from your Web site or papers. State in the conclusion explicitly what you found and why it is broadly important.

Images and graphics draw people to your poster, so make them colorful and attractive. However, such visuals should be in balance with text, so that one does not overpower the other. Arrange visuals so that they serve both as information sources and to break up text.

Clearly label the elements of a graph, image, or diagram, so people can tell what they are looking at without having to refer to the caption. For example, do not use number or letter labels with a key. A graphic's caption should not just label the graphic but also tell its conclusion. Do not number figures, since their order should be self-evident from the layout.

Positively Present Your Poster

Displaying your poster at a meeting is only the beginning of your poster communications. Consider the poster as only a jumping-off point for discussion with viewers.

To aid that discussion, create two concise verbal explanations—one for a person in your field, and one for someone not in your field. Both explanations should refer to the graphics, but not just recite or read your text.

To give your audience takeaway information, prepare a summary handout and/or reprints, as well as business cards with your Web site URL.

As you talk to your visitors, ask for questions and note them. They will help you not only improve your poster presentation but all your research explanations. Also, note the types of people lingering at your poster. If they are only your competitors, you are not attracting a broad enough audience.

6

Write Clear Research Explanations

Good writing is a high and intricate art, and it would take more than a chapter to teach it thoroughly. However, by observing some basic rules of good writing, you can improve both your professional and lay-level communication. This chapter's guidelines apply to both lay and professional writing; and see chapter 16 for good journalistic practices specific to lay-level writing.

At first you might feel uncomfortable observing these writing rules, since unfortunately your prose has likely become "infested" with the pitiable practices behind much technical writing. It will be hard to exterminate editorial bugs such as cumbersome phrases, rambling sentences, passive verbs, and dull words.

Editorially "delousing" your prose also will be tough because writing is such a very personal extension of one's ego, and it is never easy to face one's shortcomings. What's more, you might feel stripped editorially bare because dull academic prose serves a protective function, wrote Patricia Nelson Limerick, her tongue slightly in cheek, in the *New York Times Book Review*:

> Professors are often shy, timid and fearful people, and under those circumstances, dull difficult prose can function as a kind of protective camouflage.... No one can attack your position, say you are wrong or even raise questions about the accuracy of what you have said, if they

cannot tell what you have said. In those terms, awful, indecipherable prose is its own form of armor, protecting the fragile, sensitive thoughts of timid souls.

To start improving your writing, read the classic, concise guide to good writing, *The Elements of Style*, by William Strunk, Jr., and E. B. White.

Use Thrifty Words

Academic writers too often prefer "expensive" words over "thrifty" ones. Longer, expensive words seem scholarly, and you might believe they lend credibility and authority to your prose. However, overstuffed verbiage is expensive because it takes more time and effort to read and frustrates and loses readers.

Navigating a sentence jammed with expensive words is like stumbling through a stream littered with pesky rocks that stub the toe and trip the wader. Much more pleasant is ambling through a sentence stream on the smooth sand of thrifty words. The following are some thrifty words that can readily substitute for expensive ones.

"Expensive" words	"Thrifty" words
abbreviate	shorten
accelerate	speed up
accompany	go with
accomplish	carry out, do
accumulate	gather
accurate	correct, exact, right
achieve	make, do
acquire	gain, get
activate	start
additional	more, added, other
adjacent	next to
administer	give, inject
advantageous	helpful
alteration	change
anticipate	expect
a number of	some, approximately, about
apparent	clear, plain
as a means of	to
assistance	help
as to whether	whether

at the present time	now
by means of	by, with, using
category	class, group
component	piece
concerning	about, on
conclude	close, end
consequently	so
constitutes	is, forms, makes up
construct	build
continue	keep on
contribute	give
delete	cut, drop
demonstrate	show
determine	decide, find
discontinue	drop, stop
disseminate	issue, mail, send out
due to the fact that	since, because
during the time that	while
effect	make
eliminate	cut, drop
elucidate	clarify, explain, show
encounter	meet
endeavor	try
establish	prove, show
evaluate	check, rate, test
evident	clear
expertise	skill
facilitate	aid, enable, help
finalize	complete, finish
following	after
for a period of	for
for the purpose of	because, to
however	but
identical	same
immediately	at once, now
implement	carry out, do
in accordance with	by, following, under
in addition to	also, besides, to
in an effort to	to
inasmuch as	since

in conjunction with	with
incorporate	blend, join, merge
indicate	show, write down
indication	sign
initial	first
initiate	start
in lieu of	instead of
in order that	for, so
in order to	to
in the course of	during, in
is able to	can
it is essential	must
join together	join
limited number	few
locate	find
location	place, scene, site
magnitude	size
maintain	keep, support
majority	greatest, longest, most
modify	change
most of the time	usually
necessitate	cause, need
numerous	many, most
objective	aim, goal
obtain	get
optimum	best
perform	do
permit	let
pertaining	about, on
point in time	time, now, when
portion	part
position	place, put
prepared	ready
previous	earlier, past
previously	before
prior to	before
proceed	do, go on, try
proficiency	skill
proposes to	means to, suggests

provided that	if
recapitulate	sum up
reduce	cut
regarding	about, of, on
relating to	about, on
remain	stay
remainder	rest
require	must, need
requirement	need
retain	keep
review	check, go over
selection	choice
similar to	like
solicit	ask for
state	say
subsequent	later, next
subsequently	after, then
substantial	large, real, strong
sufficient	enough
terminate	end, stop
therefore	so
therein	there
time period	time, period
utilize	use
whenever	when
with reference to	about
with the exception of	except for

Besides substituting thriftier words, you can tighten prose by eliminating unnecessary "padding" words, such as

- in size
- is pictured
- it is interesting that
- month of
- on a _____ basis
- special, as in "special instrument"
- take steps to
- the process of
- type of

Also, tighten your prose by such substitutions as

- "developing" for "the development of"
- "organizing" for "the organization of"
- "applying" for "the application of"

Using precisely the right word will also improve your writing. As Mark Twain wrote, "The difference between the almost right word and the right word is... the difference between the lightning bug and the lightning."

Also, seek out not only the right word, but the most vivid, compelling word. Use the thesaurus bundled with your word-processing program or excellent online thesauri such as *Thesaurus.com*. Also, the Thinkmap Visual Thesaurus is useful because it displays web-like maps of words and their conceptual relationships.

Make Sentences Sing

Keep average sentence length short. Research on reading has shown that as sentence length increases, text comprehension, even for educated people, drops drastically. An American Press Institute study found that readers typically understand 100 percent of the information in 8-word sentences, 90 percent in 15-word sentences, but only 50 percent in 28-word sentences. However, a string of short sentences tends to slow reading, so mix up your sentence length. Use short sentences to give ideas punch and longer sentences when needed for a more complex idea. Like this.

Most word processors offer readability tools that enable you to check the average sentence length and other measures of reading ease. In Microsoft Word, to switch on the readability statistics:

- Click on the Microsoft Office button.
- Click on *Word Options.*
- Click on *Proofing.*
- Under *When correcting spelling and grammar in Word*, make sure *Check grammar with spelling* is selected.
- Select the *Show readability statistics* check box.

When Microsoft Word finishes checking spelling and grammar, it displays average words per sentence and two measures of the document's reading level—the Flesch Reading Ease score and the Flesch-Kincaid Grade Level. The higher the Flesch Reading Ease score, the more understandable your piece. The scale runs from 0 to 100. Simple prose such as comic books scores in the

90s, and such scholarly works as the *Harvard Law Review* score in the 30s. Popular journalistic writing ranges around the 50s and 60s. Writing about science, medicine, and engineering tends to score a bit lower, because of the complexity of the concepts and terms. The Flesch-Kincaid Grade Level score, of course, indicates the educational level most readers would need to understand your piece.

You might have trouble writing shorter sentences because the scientific communication culture has instilled in you that longer sentences are more authoritative. For example, in one telling experiment, communication researcher J. Scott Armstrong asked faculty members to rate the prestige of passages from management journals. While Armstrong held constant the content of the passages, some were written in long, complex sentences with unnecessary words and others were more readable with shorter words. Armstrong found that the professors rated the verbose passages as higher in research competence.

To create readable sentences, also take into account their "hammock" structure. That is, the most important ideas in your sentences should come at the beginning or the end. For example, see how less effective the following version of the previous sentence is, because it puts the important thought in the middle: *You should put at the beginning or the end the most important ideas in your sentences.*

Write Actively

Although researchers widely use passive voice in professional writing, avoid it wherever you can. As William Germano wrote in an eloquent essay in the *Chronicle of Higher Education*, passive voice conceals responsibility for action:

> The active voice should be a kind of scholarly credo: I did research, I drew conclusions, I found this out. That's rarely what we get. How much more often do we read that research is conducted, conclusions are drawn, findings are found out? I sometimes imagine a scholar sitting down with a great idea, then staring at his laptop and exclaiming "Are you crazy? You can't say that—" and clicking the toolbar to call up Active-Voice-Replace, instantly turning every "I found" into "It was discovered."...It's particularly critical for young scholars to understand that all this bother about the passive voice isn't simply a matter of making sentences lively, peppy, or more engaging. Yes, the active voice is stronger. Readers listen more attentively because they can hear another human trying to engage their attention. But for scholars, the active-passive conundrum should be so much more. The active voice says

> "I have something to say, and I'm going to say it. If I'm wrong, argue with me in print. But take me at my worth."

However, as Germano—vice president and publishing director at the academic publisher Routledge—points out, passive voice does have some uses.

> Of course, it's important to draw a distinction between writing with the passive voice and writing in the passive voice. In the first case, the writer uses the passive when it's necessary. In control of her prose, she enjoys the way the passive voice lends variety to her sentences, yet she remains the boss in her own paragraphs. On the other hand, someone who writes in the passive hopes no one will notice that she's there. The passive is a cozy place to hide.

Write for the "Reading Eye"

Your writing should not only engage the mind but be easy for the "reading eye." To explain: the eye does not scan text smoothly, but fixes on a word or phrase and then makes rapid jumps, called saccades, to the next. During a fixation, the reader's peripheral vision registers the next word or phrase. Thus, people read chunks of words and phrases to comprehend text, rather than serially recognizing letters. This "parallel letter recognition" model of reading holds that readers simultaneously recognize all the letters within a word or phrase and use those letters to recognize it.

The plraaell lteter rgctenooiin mdeol eplxnais why you can raed this stenecne filary wlel, even tguhohg olny the fsirt and lsat leertts of the lnoegr wdros are in oerdr.

Given this perceptual machinery, you can make your writing more readable by providing the eye with visual landmarks in the form of distinctively *shaped* words and phrases. For example, I sometimes revise a sentence whose words have too many planar letters—a, c, e, m, n, o, r, s, u, v, w, x, z—to substitute words with letters featuring the visual landmarks of ascenders and descenders: b, d, f, g, h, i, j, k, l, p, q, t, y.

The eye's need for visual landmarks helps explain why passive verbs—is, are, was, were—are so ineffectual. They fail to provide landmarks for the reading eye because they are short words with no visually distinctive letters. Thus, the reading eye easily overlooks them, and they fail to aid the reading process.

Writing for the reading eye also means creating layouts that offer visual landmarks in the form of white space, paragraphing, drop caps, pull quotes, and so forth. For example, in a well-designed layout, a paragraph is usually no longer than the width of a column. The resulting chunk of text is easier to read. As a writer, you will not usually control the layout of your articles. But if you become an editor, listen to your designer and appreciate the importance of good layout to the reading process. Perhaps you may someday even help remedy the miserable state of scientific journal design.

7

Build a Quality Web Site

Most researchers' Web sites exhibit such appalling design and content, I sometimes wonder if they make their site's shabbiness a perverse point of pride—like adolescents wearing tattered jeans. Such disregard for your Web site is risky for your work because it will be exposed to a vast audience and constitutes a central information source for important audiences. Your prospective students, donors, colleagues, university administrators, media, and relatives will all explore your Web site to find out about you and your work. And more than half of Americans choose the Internet as their main information source about scientific issues—and that includes your research, according to the National Science Board's *Science and Engineering Indicators 2008*. The Internet is second only to television as the primary source of information about science and technology, and its influence continues to grow.

Your institution or department might provide professional design templates and other services to help you create and maintain your site. For example, the Duke neurobiology department has long committed to making its Web site an exemplary one that portrays the department's excellence. And the Howard Hughes Medical Institute (HHMI) shows how a larger organization recognizes the benefits of devoting resources not just to quality design, but to quality content.

Whether you have institutional support or not, invest in designing and maintaining your own quality site. The effort will pay enormous dividends, because a quality Web site:

- **Establishes your "brand" reputation.** Thinking of yourself as a brand like canned tuna or laundry detergent might seem a bit distasteful. But like it or not, your name and your institution are brands. When your audiences see your name or that of your center or institution, they should recognize it as a quality brand name because of the reputation you have built. Of course, your reputation depends most critically on the quality and significance of your work. However, communicating that work in a professional, accessible Web site portrays that quality to fellow researchers and to lay audiences. Chapter 15 covers other important communication outlets—including e-mail newsletters, blogs, and webinars.
- **Tells your research story your way.** Your Web site is an excellent way to communicate the larger picture of your work, as well as the detailed pieces that form that picture. Your papers, grant proposals, and seminars, as exhaustive as they are, inadequately portray your work to the outside world. They offer little accessible information about your overall philosophy and research strategy, and implications of your work.
- **Gives reporters a head start on understanding your work.** "Every researcher should have a fantastic Web site that tells what the lab is doing, as well as all the articles as pdf files," says *New York Times* science reporter Sandra Blakeslee. "And it should be updated regularly. It is just a gold mine. I will read those papers, so I can ask better questions based on them. If I don't have that material it is much harder to get started and it takes more of the researcher's time."
- **Integrates people into your research community.** Your Web site can uniquely present the community of your laboratory to your own people, as well as the rest of the world. It can tell both the personal and professional stories of you and your researchers, making your laboratory and its work more accessible. For example, when HHMI names new investigators, it develops extensive bios and research descriptions about them to post, along with professionally done portraits, on its Web site. "It is one way that we help bring new investigators into this large family of scientists," says HHMI public information officer (PIO)Jim Keeley. "It helps in telling the stories of the investigators, why they were selected, why they are interested in science, why we are excited about their science."
- **Takes your work global.** Your Web site gives you instantaneous global visibility. It constitutes the main searchable source of information on

you and your work. And a compelling, news-filled Web site can attract an audience far beyond people in your field and your institution.

- **Provides longevity.** Web content is surprisingly long-lived. It does not depend on being reprinted, and it does not get lost in a drawer, but is literally at your audience's fingertips. For example, when HHMI measured the audience for its news service, it discovered that some of the highest-ranked stories were years old. "We still don't know the reason for that finding," says Keeley. "The only thing we can think is that if it was a good story at the time we did it, it's still a good story three years later, and there is an audience for it."
- **Is always current.** Unlike print material, you can instantly update your Web site, and with such services as RSS feeds, your audiences can immediately learn of those updates. RSS, which stands for "really simple syndication," consists of software that enables people to subscribe to news feeds that automatically send new content to RSS reader software, called "aggregators."
- **Establishes you as a good institutional citizen.** Since the Web is just that—a *web*—your Web site is linked to the rest of your organization. So creating a quality Web site enhances not only the quality image of your work but also the image of your institution. The leaders of your institution will certainly appreciate your good institutional citizenship.
- **Makes you a better communicator.** The Web is a demanding medium in the editorial discipline it requires to produce effective content. Web text must be tight and focused, giving special attention to design and layout to make content accessible and visually attractive. Many journalists and PIOs say the discipline of writing for the Web has made them better print writers. As a researcher, you will likely find that the process of developing a quality Web site will enhance your ability to communicate even more profoundly.

Plan Your Web Site

A first step in creating your Web site is to understand your institution's policies on content, style, liability, and other issues. They may differ among institutions and types. For example, universities are often laissez faire about their faculty's Web sites, requiring only that content not be libelous or obscene. However, corporations and federal laboratories may tightly restrict content and design. Your institution's webmaster can brief you on both formal policies and informal practices and direct you to resources and exemplary sites.

Do Discovery First!

Once you understand the Web policies, but before you hire a designer and plunge into creating your Web site, begin with a "discovery" process to define your site's audiences, their needs, and how your site can meet those needs. Conduct this discovery process yourself or hire professional help. Either way, form a Web advisory group that includes colleagues and outside representatives of your lay audiences. "Outsiders can be valuable because they can ask the 'dumb' questions," says Web content strategist Merry Bruns. "They can ask questions about the site and its content that the researchers might not have thought of themselves."

Begin your discovery process by considering what your laboratory and your research field dictate in a Web site. You might only need a basic text-oriented site, but if explaining your research requires visuals or animations, you should invest in a more sophisticated Web site. For example, Duke neurobiologist Dale Purves and his colleagues explore the nature of visual processing. So, the Purves laboratory Web site at *PurvesLab.net* features extensive use of interactive animations.

Similarly, the Hubble Space Telescope site at *HubbleSite.org* needs to properly display the images from the telescope, to engage and educate visitors. This stunning site offers a wealth of such visuals, even enabling users to download them as computer wallpaper. What's more, the site's content takes advantage of the "teachable moment" presented by the images to lure people into astronomy. The NASA Web site at *NASA.gov* and its science site at *NASAScience.NASA.gov* are also tours de force of Web site design and content. They contribute greatly to NASA's research and education missions, not to mention its political aims of garnering support for those missions.

Your resources, however, are likely far smaller than NASA's, and your work might not involve elaborate graphics or animation. Even so, you can still use the design and content guidelines below to create a site that effectively lures users into your research.

In beginning your discovery process, become a Web "spy"—exploring sites like yours to assess their design and content. Users will invariably compare your site to those, and you will want it to compare favorably. Continue to spy even after your site is live, to gather ideas and links to apply to your own site to make it even more valuable to your audiences. Also, learn about Web usability. Good sources for usability information are *WebPagesThatSuck.com* and *UseIt.com*, operated by Web design guru Jakob Nielsen, as well as Nielsen's books, particularly his 2006 *Prioritizing Web Usability* coauthored with Hoa Loranger.

After your advisory group has explored the comparable sites, work with them to define your audiences, what they will want from your site, and what information you can provide. If possible, survey users about what they want in a site such as yours. From these data, create a matrix that lists audiences along one axis and the type of information or functionality they want along the other. This matrix will reveal what content or functionality to feature most prominently and devote the most effort to creating. For example, if potential users of shared equipment constitute an important audience, you would feature your instrumentation and its capabilities more prominently. Or, if you seek to educate donors about your work, or to attract patients to clinical trials, you would highlight information about how your research contributes to treating your target disease. Also, to make your site useful for those audiences, your site would feature extensive background about the disease and links to outside resources.

If your analysis reveals a significant need, do not hesitate to fill it, even though the content might not be conventional on a site like yours. For example, while most visitors might want to read your research papers, a significant number might also enjoy that clever science game you created for your child's science class.

Also think "negatively," advises Bruns. Ask yourself why audiences might avoid exploring your site. The negatives will inform your strategy just as much as understanding the positives of your site's attractions, she says. For example, is your work's theoretical nature likely to intimidate nonscientists you want to reach? If so, create a special lay-level section that lures the lay user by offering engaging visuals, stories, metaphors, and other content.

Next, use your audience analysis and content matrix to map your site's architecture and navigation. This will include the menu categories, what Web pages they link to, and how the site is organized. One alternative to the usual category-based navigation, says Bruns, is audience-based navigation (e.g., "For researchers," "For students"). "This works if your audience base is divergent," she says. "However, if your audience base is very close together it doesn't. For example, media and scientists are far enough apart in their interests, but patients and consumers are too close together." However, says Bruns, a site should never have just audience-based navigation. "It's far too limiting, because you get what I call the 'Gee-I-wonder-what-the-other-guy-has-got?' syndrome. Also, people might fit in multiple audiences, and how do they choose?"

Duke communication director David Jarmul warns in particular against an institution-centered site organization. "So many sites at universities are arranged through the eyes of the originating unit, with incredible detail that reflects the organization chart, rather than the needs of the users," he says. The site *WebPagesThatSuck.com* offers an excellent online resource for learning about good, and bad, Web organization and design.

One feature of good organization is that the site's basic navigation links are "a rock" on the Web site, emphasizes Bruns. "The navigation for the main sections of the site should be on every page, and they don't move ever," she says. "A user will come into your Web page from anywhere. Google can send them to any page, or they can be bouncing around your site ending up not knowing where they are. They can always go back to that rock of your basic navigation and find their way back."

Good navigation uses "breadcrumbs," which is a visual line of hierarchical links that show a user exactly where he or she is in the navigation structure. Clicking on any breadcrumb link allows the user to backtrack anywhere in that hierarchy.

Also, creating a site map during development enables you to visualize the pathways users can take through your site. This map is basically a block diagram of the site showing how pages link to one another. You can sketch the site map on paper, but to make it easily shareable, produce an image of it using programs such as Adobe Illustrator or Photoshop. Circulate this site map among your advisors and incorporate their feedback.

Once you have your basic navigation structure, test it by creating a "wireframe"—basically a simple working Web site with the link structure but no design or content. Ask not only your advisory group but also others to test the wireframe, says Bruns. "The key thing is not to let anybody that works with you participate in the test. It has to be people who have no idea who you are. They're going to try out your wireframe and at some points just say 'Huh?' That is what you want to hear." The "huhs" will be invaluable clues to how you need to improve your navigation, says Bruns.

In creating large sites, professional firms usually test usability with many groups. However, if you are only developing your laboratory site, ask an in-house Web development group to do more modest testing. As a last resort, conduct the testing yourself, dragooning anybody who is willing to try out your wireframe and give you feedback. In such cases, Bruns's advice is to "shut up and listen." Do not try to explain your site; its functionality should be self-evident.

Get a Memorable URL

Both the Purves and NASA sites illustrate the value of a short, memorable URL. The Purves lab site is, logically, *PurvesLab.net*, the Hubble site is *HubbleSite.org*, and the NASA science site is *NASAScience.NASA.gov*. Similarly, the Genetics and Public Policy Center at Johns Hopkins University uses the simple URL *DNAPolicy.org*.

Your Web address need not be a long hard-to-remember string that includes your institution's name. You can use a descriptive URL such as those above, because any URL can be redirected to your institutional URL. And you can still get the advantage of your institution's imprimatur by including it in the URL, for example, *smithlabatharvard.org*. Incidentally, there is a problem with this URL as written: it can be misread as smith-la-bath-arvard. While this URL is only mildly awkward, there are other outright embarrassing URL gaffes. For example, *whorepresents.com* is the Web site of a service that lists Hollywood agents, and *expertsexchange.com* is the site for finding computer mavens. So, pay attention to possible misreading of your URL.

Capitalizing your URL's key words in print, as I have done previously, will make it more readable, as in *ExplainingResearch.com* or *SmithLabatHarvard.org*. Even if users do capitalize when they type in your URL, it will still work. But avoid hyphens and underscore marks in your URLs. They are harder to remember, are often overlooked by users typing in your URL, and are harder to recite out loud, as in talks or on the radio.

To make your site even more accessible, you can also have multiple special-purpose URLs that you redirect either to your home page or to specialized pages. For example, the URL *WorkingWithPIOs.com* redirects to the page of the *ExplainingResearch.com* Web site on working with PIOs. Also, register popular misspellings of your URL—for example, both *DennisLab.org* and *DenisLab.org*—and redirect them to your site. And register any domain name that you might even consider using. After all, registering commercial domain names is very cheap, and registering domains within your institution is likely free.

Design a User-Friendly Web Site

After your usability testing, begin your design process by collecting sites whose design and functionality you like. Think about what features from these sites you want to incorporate into yours. Do you like the menu layout? Their look and feel? The use of the logo or graphics? Remember, your design need not be elaborate or expensive to be effective, says Bruns. "If you are going to put your money anywhere, put it up front in the content strategy and navigation," she says. "Believe it or not, you can have a very simply designed site, and it can be the most usable and useful one on the planet."

A good example is the very functional but simple Web site of the Genetics and Public Policy Center, *DNAPolicy.org*. The site needs no elaborate graphics or design because it aims to be an authoritative, informative policy site. Rather, the site uses clear writing and well-designed navigation to communicate its message. I like to believe that *ExplainingResearch.com* falls into the category of simple-but-useful sites, as well.

If you hire a designer, make sure the candidates for the design work understand how to portray such complex topics as your research. They should also understand the "Web culture" of the organization to which you belong—university, corporation, government laboratory, and so on. Candidates also should understand your institution's design standards and rules for posting content, as well as unwritten rules of the culture. The designer should be a strategist, not just an artist, says Bruns. "The kind of questions you want to ask are, not so much 'How did you do that pretty picture?'" she says. "Rather, ask them to show you a particular problem one of their clients had with navigation, and how they solved it with design. A good designer will come up with ideas you never thought of."

If you cannot afford a professional designer and your institution offers no design help, you can create your own site using tools offered by Web hosting and design companies. See the online resource section at *ExplainingResearch.com* for links to sites that offer design templates and tools that do not require you to know HTML or Web design programs.

Before you develop your site, though, make sure there are no institutional rules against creating your own site and/or hosting it outside the institution. If there are no strictures, explore the collections of design templates to determine whether one has the look and feel you want. Ask potential users and colleagues whether the templates you like present the right image for your laboratory. Finally, determine whether the hosting company you want to use for your site offers the site features—such as calendars, photo galleries, and e-mail newsletters—that you want.

Whether you hire a designer or create your own site, follow good basic design principles. Here are a few of the most important:

- **Go for a professional look and feel.** You want your Web site to portray your laboratory as a place of professionalism. So, go for a clean, uncluttered design. Avoid a complex background and too many design elements, links, or images. Also, choose handsome, complementary colors arranged to guide the viewer's eye.
- **Make your home page a showcase.** Just as people do judge a book by its cover, they also judge a Web site by the first screen. "On the Web all the reader sees at first is whatever shows up on the top part of their browser window," says Bruns. "They are not going to scroll at first, so that first screen is the amount of room you have to pull them into whatever it is you want to show them." Web usability studies reveal that you have about 1/50th of a second to draw people into your home page. So, your design should put the most important and engaging links and content in that first screen.
- **Attract the eye.** Design your home page with white space and a balance of text and interesting photos to attract the eye and thus engage your users.

When arranging content, use a hierarchy that reflects how the eye scans a Web page. Eyetracking studies reveal that people scan a Web page in the following order:

1. Top left
2. Top middle
3. Down the middle column
4. Left column below the top
5. Top right
6. Down the right column.

- **Help the reading eye.** The computer screen is harder to read than the printed page, since its resolution is lower. So, help the reading eye by using headlines, topic subheads, bullets, and other layout features. Also, use the hypertext capabilities of the Web to their fullest. Rather than displaying all your text, charts, and graphs on a main page, make it a "backbone" of your document, with clickable links to additional references, charts, or graphs.

Write for the Web

Writing for the Web is very different than writing research papers, requiring different organization and style. Here are some organizational techniques for making Web text accessible, from Web usability expert Chris Nodder:

- Start with the conclusion you want users to reach.
- For longer articles, write abstracts or summaries.
- Tell readers what questions they can expect an article to answer.
- "Chunk" text into paragraphs of about seven lines, since Web usability studies show that users have difficulty taking in longer paragraphs.
- Make titles, headlines, and subheads informational, not just labels.
- Make the clickable text of a link long enough for easy mouse-clicking.
- Use relevant key words in the text that will enhance the link's visibility to search engines.

The same layout features that make print easy to read—such as page numbers and a table of contents—also enhance accessibility, Web readability studies have found.

Use images that contribute to the story and are not just decorative. Merely decorative graphics are barriers to understanding content, according to a Web eyetracking study by Jakob Nielsen and Kara Pernice Coyne. If you do not have a relevant image, use white space, they advise. The study also found that readers are more interested in looking at "approachable" people, not models, and they like people who are smiling and looking at the camera.

Tightly written text is also critical to communication, the studies have found. For more detail on Web writing, see the book *Hot Text: Web Writing That Works* by Lisa and Jonathan Price and their site *WebWritingThatWorks.com.* Also, Bruns's Web site at *ScienceSitesCom.com* includes a comprehensive Web editor tool kit and bibliography.

Even Web pages that explain the technical details of your work should be as concise as possible. Such concise writing does not mean leaving out critical details, says Bruns. "We are not talking about dumbing down your work. The Web is a new medium, and it requires that information be structured differently from print—not to change its meaning but to make it stronger, clearer and more focused on your key points," she says. Bruns advises that the Web text aimed at colleagues should read like what you would say to them at a conference. However, she emphasizes, keep in mind that the same text also will be read by students or other laypeople. So use less jargon and fewer acronyms. And when you must use technical terms, link each term to a glossary entry.

If you doubt your ability to write for the Web, consider hiring a professional writer, advises University of Wisconsin PIO Terry Devitt, a founder of The Why Files. "A professional communicator will bring to the table knowledge of how people absorb information," he says. "And, while scientists may know their specialized audiences and how to communicate with them, they don't necessarily know how to communicate with the rest of the world. A professional communicator knows what interests people and how to encode information in a language that makes it accessible to broad audiences."

Feature "Heroin Content"

Despite a long-standing myth that people do not read online, new studies show that people will read substantive Web text, including research explanations. For example, a 2007 study by the Poynter Institute found that people will, in fact, read a larger percentage of a story text online than they will in print. The study showed readers were helped to understand content by alternative story forms, such as Q&As, timelines, short sidebars, and lists—substantive information that Vincent Flanders, who runs *WebPagesThatSuck.com,* dubs "heroin content." The heroin content on your site might include the following:

- **Complete contact, location information.** This includes e-mail addresses and phone numbers of laboratory personnel, its physical address, travel directions, links to maps of your institution, and Google or MapQuest maps of the area. You can also create a personalized Google map of your laboratory and its surroundings that can include your own text, photos, and videos.
- **Complete bios.** Go beyond the usual CVs or dry, canned bios for your site. Users will find comprehensive, personal biographical sketches much more interesting and useful. For example, HHMI develops professionally written biographical sketches for its investigators. "These are different than what investigators might do for themselves," says Keeley. "They are more of a narrative that talks about the researchers' own interest in science, how it developed, and what questions they are interested in asking. It was a significant investment in time and effort to generate the bios, but in the long run we are much better off for having them. For example, they provide great background on investigators when they get requests to speak and when they have media interviews."
- **Engaging research descriptions.** Similarly, descriptions of your research should be more than the dust-dry text lifted from your latest funding proposal. HHMI tries to make research descriptions engaging, says Keeley:

  > We ask investigators to make their descriptions as accessible to a lay audience as possible for lots of different reasons. One is you are not writing it for your peer group, for researchers in your own field. They will be read by the public, students, teachers, reviewers, people looking for review panel members, people who work for companies looking for collaborations, and by politicians and public policy people. They are not asked to write about experiments they haven't published yet, but they are asked to write very broadly about their plans.

- **A feedback form.** A form that enables your audiences to make comments, ask questions, and offer ideas will make your site more engaging. The feedback will prove an invaluable source of content for your site and ideas for improving it. Of course, having such a feedback capability means you must respond to audience members.
- **FAQs.** Not only will your audience find a set of frequently asked questions about your research interesting, but it will save you time.

A FAQ allows people to get answers to many questions themselves, and you and colleagues can lift standard answers to such questions for your own letters, memos, and other communications.

- **Q&As.** Compile the most popular questions from your feedback form into Q&As, which are more personal than a FAQ. They present you as an expert answering questions in an accessible and interesting way.
- **Comprehensive facilities information.** Include information on your facilities and research capabilities that colleagues, prospective collaborators, prospective students, and so on, will want to know. Consider developing a photo tour of your laboratory, which you can create with image gallery software, as outlined in chapter 12 on photography. Or, use a photo-sharing site such as Flickr to create a photo tour of a larger research complex.
- **Accessible scientific papers.** Post as many of your scientific papers as possible, at least as pdf documents, but if possible as HTML documents with links to references and other supporting material. Be assertive with scientific publishers about being allowed to post your papers. That assertiveness will pay off in enabling important audiences to access your work. If you cannot get posting permission, at least have a link to the subscription-only journal site or to a pay-per-view site such as *ScienceDirect.com*. Most science journalists have access to such sites. "It drives me crazy when someone will have only citations but no pdfs," says Blakeslee. "I have to hunt the papers down. If you have a hundred papers, though, you don't have to put them all up, but at least the major papers, and certainly your review papers." Blakeslee recommends even having a special media section with links to papers in which reporters might be interested.
- **Technical tips, reviews, how-to guides.** Offer tips on laboratory techniques or instrument operation from your experience. How-to guides can either be professional or lay oriented. For example, an astronomer could offer tips on new telescope instruments, or an automotive engineer tips on car care. Similarly, you can review both scientific equipment and consumer products related to your discipline. Such reviews can originate in a blog, as discussed in chapter 15.
- **News section.** Include links to internal and external media coverage of your work. Caution: many media stories, although public at first, may be transferred to a paid archive. And other stories will always be subscription-only. So, obtain permission to post the story text or a pdf on your site. You do not need permission to post a link to the story. If there is a significant gap in news coverage of your work, consider preparing your

own Web-only news story to post on your site. However, coordinate with your PIO to make sure institutional policy allows such news stories. Some institutions frown on researchers doing such stories, even if they are only posted in your laboratory or departmental site. On the other hand, your PIO might be prompted to do a story on the work or to link to your story.

- **Media kit.** If you are a "public scientist," as discussed in chapter 27, consider creating a media kit on your site. It should contain publication-quality photos of you, a brief bio, newsworthy topics you are willing to discuss, suggested interview questions, links to books and popular articles on your work, and testimonials from producers on your performance as a guest. Some of these links, such as the news article links, can be the same as those under other sections.
- **Human-interest stories.** Besides interesting bios, human-interest stories about your laboratory will make your site more interesting and accessible. It also will help dispel the myth of research as a dry, bloodless endeavor and researchers as banal people. For example, the Duke Pratt School of Engineering posts engaging profiles of its engineers that portray engineering as the dynamic enterprise it really is, says school PIO Deborah Hill. "People still have very old ideas about what engineering is, and I didn't see anything on our Web site that could counter that," she says. "We weren't making the most of some of the very exciting things that our engineers are doing that are counter to what people think engineers do. Engineers don't just build bridges and roads; they also build artificial hips and engineer medicines to go into tumor cells." So, Hill and her colleagues produced profiles "so that students could see that engineers are not guys with greasy hair, black glasses, and pen protectors. They are dynamic, exciting men and women who go salsa dancing and create technology that will help people."
- **Fun stuff!** Adding humor and personal flair to your site will not detract from its credibility. "Don't take your stuff so seriously that you can't have fun with it and make other people see the fun in science," says Devitt. "By doing a bad job of portraying the fun of science, scientists have unfortunately done a good job of excluding a lot of people from science." Consider offering cartoons, games, or personal anecdotes on your site. Of course, the design of your site should segregate the fun stuff from your serious material.

Make Your Site a Go-To Resource

Your "heroin content" can include comprehensive background on your field, besides the information on your own work. Links to general reference content attracts users to your site, prompting them to explore your own work. And, such background will cause users to perceive your site as more authoritative and useful and to spend more time there. You can also link to such general background material from your technical content, enabling lay-level readers to understand technical concepts. As a prime example, the Harvard Medical School site includes not only articles about Harvard research but also comprehensive health information and subscriptions to the Harvard health newsletters.

It is easy to identify such broader resources. You likely already know of links to good information sources in your own field. And the online resource section at *ExplainingResearch.com* lists a wealth of links to authoritative science and technology sites. These sites include government agencies such as NIH and NSF, media such as the National Geographic Society and *Science News*, and professional associations such as the American Astronomical Society and American Physical Society. The resource section also includes link collections, such as those on EurekAlert! from which you can copy good links. These collections include glossaries, background articles, editorials about your field, and sites that are just plain fun, like the delightful collection of chemistry videos, Kent's Video Chemical Demonstrations.

Social bookmarking sites, discussed in chapter 15, are another good source of links to articles about your field. These sites include Digg, deli.cio.us, Mixx, Reddit, StumbleUpon, and Yahoo! Buzz. Basically, users of these sites post links to articles they find interesting. Also, you can subscribe to EurekAlert's RSS feed to receive news releases, and to Google Alerts to automatically be notified by e-mail of news, Web pages, blogs, video, and discussion group postings on a designated topic.

You can also copy links from your own institutional sites. For example, at Harvard, you could link to Harvard's Research Matters site, which features articles on Harvard research. At Vanderbilt, you could link to Vanderbilt's online *Exploration* magazine, whose research stories feature handsome graphics. For a list of online university magazines to which you can link, see the University Research Magazine Association site.

Of course, link only to authoritative sources. For example, although Wikipedia is a well-known and usually accurate source of information, the article on your field may not be authoritative enough to link to. However, if you write and/or monitor the accuracy of the Wikipedia entry in your field, you could safely use it on your site. Also, Google's Knol contains articles by recognized authorities, which you can use if they are accurate.

Be Image Conscious

High-quality images are extremely important on a Web site, as confirmed by the Poynter Institute eyetracking study cited previously. The study found that users do pay considerable attention to Web site images, and that photos of people doing things attracted more attention than staged or studio photos or mug shots. Certainly, your site should not display the all-too-prevalent, amateurish mass shot of your laboratory group lined up on the lawn. Rather, feature professional-quality images depicting your work and your laboratory, as well as head-and-shoulder portraits of your laboratory personnel. And if your work generates dramatic images, by all means feature them as stand-alone shots or in a gallery.

Chapter 12 covers in detail how to work with photographers to obtain professional-quality images. However, if you have the experience and ability to create good images, by all means do so. You may also find images on the Web—both free and commercial—that can enliven your site. See the list of image sources in the online resource section, under chapter 3.

Keep Your Site Fresh

Stale or static content compromises the credibility and quality of your Web site. Regularly add new technical papers, abstracts of upcoming talks, the latest laboratory news, and other content. Also, post comments, ideas, and discussion that your audience contributes via your feedback form. To highlight this content, include a What's New section on the home page.

Besides giving your site a sense of dynamism, continual updating increases the number of visits. The effect of new content on traffic can be extraordinary. For example, the news pages on the HHMI site, which change constantly, average about six million page visits a year, representing about three million visitors. The news section is the most visited content area within the site. Also, new content triggers search engine "bots" to pay more attention to your site, raising its visibility in searches.

The need for continual updating means that your site should not consist simply of static HTML pages, which are harder to update. Rather, the site should be fed by a database, using a content management system to manage data. Content management systems, although more expensive to implement, can be easily updated by anyone, without the need for knowledge of HTML coding.

Market Your Site

If you build it—in this case your Web site—they will *not* necessarily come. You can create the most elegant Web site ever, and unless you do active marketing and outreach, you will likely not get the traffic you expect. Fortunately, marketing your Web site is not difficult. Dell offers a good online course in developing and marketing a Web site, and books such as *Web Marketing for Dummies* offer useful guides.

Here are other good tips on marketing your site:

- Include your Web site URL in every presentation, publication, and video and list it on every letter and e-mail you send. Use the full URL in e-mail messages, so that it will appear in received e-mails as a clickable link, rather than text that has to be copied and pasted into a browser address window. For example, *http://www.ExplainingResearch.com* and *www.ExplainingResearch.com* transmit as a link, while *ExplainingResearch.com* does not.
- Submit your site to Google, Yahoo!, MSN, Open Directory Project, and other search engines.
- Request other Web sites to link to yours. This is easier if you offer the kind of heroin content that makes outside links useful. These other sites might include those of your professional association, your department or center, museums, and other research-related institutions. Web sites often exchange links—that is, "I link to you if you link to me." However, search engines have caught on to this practice as a ploy to artificially increase the number of links to a Web site—a practice aimed at increasing a site's visibility in search results. Now, search engines do not count such reciprocal links in their assessment.
- Use effective "meta tags," which are descriptive words that are part of your page coding. These meta tags include:
    - An informational title, which is a short key-word-rich sentence that describes your site and appears on the top bar of your browser. So, your title should not just be "Frankenstein laboratory," but rather "Frankenstein laboratory for reanimation using lightning and surgically joined body parts."
    - A key-word-rich description, no more than about 200 characters describing your site.
    - A full complement of key words describing your research that will be recognized by search engines. Do not just use general key words, for example, *lightning* and *surgery*. Add all the words and phrases relevant to your work that people might search on, for example, *reanimation*, *monster*, *brain transplants*, *limb attachment*, or *grave robbing*. Put the most important words first. As a test, search on the key words that most describe your research area and see which sites come up first. Also, look at related sites to see what key words they

use. Review your key word list often and update with new terms as appropriate.

- Use key words wherever possible in the text links on your pages, because search engine software spiders that survey Web sites weight them more heavily in ranking your site. Thus, a text link on the Frankenstein laboratory site might be "use of carefully selected body parts" rather than just "parts."
- Include ALT tags on all your images. These are the text labels for an image. Search engines can read these tags, but they cannot decipher the content of an image itself.
- Use Web optimization tools offered by Google, Yahoo!, MSN, Ask.com, and others to enhance your site's visibility to search engines. URLs of such tools are included in the online resource section at *ExplainingResearch.com.*
- Enhance traffic to your Web site by creating pages on the popular social networking sites such as Facebook, MySpace, and Squidoo, which direct users back to your site. Chapter 15 covers such social networking sites.
- Continually track your site's traffic using such tools as Alexa.

PART III

Engaging Lay Audiences

"One thing I'll say for us, Meyer—we never stooped to popularizing science."

8

Forge Your Research Communications Strategy

Fortunately, the elitist attitude of the scientists in the cartoon that introduces this section is largely defunct, but its remnants still linger on. Some researchers still pejoratively call adapting their explanations for lay audiences "dumbing down"—revealing themselves as communication illiterates who do not appreciate the validity or value of lay-level communications. This chapter and the others in this section aim to show why explaining your research to lay audiences is critical to your career and to offer the tools to make those explanations effective.

Chapter 2, on planning your research strategy, will get you started considering such conceptual communication-related issues as assessing your own goals and interests and adopting the most productive attitudes. This chapter covers the practical steps to creating that communications strategy to enhance your research and your career.

First, Protect Your Publication

Of course, your prime objective is to ensure that your scientific publication is not compromised by news releases, Web sites, or other lay-level communications. For many journals, such public communications are not a problem. For example, the American Geophysical Union and the American Physical Society place no restrictions on publicizing your papers as soon as their journals accept them. And even before a paper is accepted, they allow full and free discussion of your findings with media and any other audiences.

However, many prominent journals observe the "Ingelfinger rule," which holds that they will not consider a manuscript for publication if its substance has been submitted or reported elsewhere, including in the media. This policy was promulgated in 1969 by Franz J. Ingelfinger, then editor of the *New England Journal of Medicine*. Besides the *NEJM*, other journals that observe the rule include the American Chemical Society journals, Cell Press journals, *JAMA*, *Nature* journals, *PNAS*, and *Science*. What's more, after those journals accept a paper, they place on the paper a specific embargo date and time, before which media may not publish stories on the findings. The AAAS media policy for *Science* is typical of embargoed journals. As AAAS public programs director Ginger Pinholster explains, *Science*

> does not seek to inhibit scholarly exchange of information; and so authors with pending papers are perfectly free to present their findings at scientific conferences as they normally would. All we ask is that they not participate in proactive publicity related to those conferences—such as issuing news releases and participating in press conferences. And, we ask them not to describe submitted work as pending at *Science*, because if it hasn't been accepted, we can't provide our own imprimatur just yet.

Such proactive publicity even includes helping in-house writers produce preembargo content for newsletters or local newspapers, points out Pinholster. In the age of the Internet, every publication is national, even international. A "local" story on your work, even in your in-house newsletter, might compromise scientific publication by finding its way onto Web news sites or being used as the basis of a national reporter's story. And even if such a story does remain in-house, it could constitute the kind of proactive publicity that a journal editor would see as precluding publication in a journal such as *Science* or *Nature*. Similarly, posting results on your laboratory Web site, unless password protected, constitutes public dissemination of those results. However, it is permissible to contact your public information officer (PIO) in advance to begin planning embargoed releases, emphasizes Pinholster.

Be preemptive in protecting your scientific publication by anticipating which journal you might submit to. For example, if you might submit a paper to *JAMA*, *Nature*, or *Science*, be circumspect about public discussion of your work. But if you will publish in journals that have no embargo policy, you can feel free to post findings on your Web site or on public preprint servers or to discuss them with the media. You should be aware that there is a sharp disagreement among journalists, PIOs, and journal editors over such embargo policies, which is covered in the "Working with PIOs" section at *ExplainingResearch.com*.

Learn Your Institution's Communication Policies

Your institution's communication machinery and policies will greatly influence how you reach out to broad audiences to explain your work. These policies include media policies, internal communications, and "branding" policies.

Media Policies

Different organizations may have very different policies about whether and how you can talk to the media. Corporations almost always require that reporters contact you only through your PIO and that the PIO be present during interviews to ensure against release of proprietary information. Medical centers usually allow direct contact, but they operate under federal privacy regulations requiring that a PIO escort reporters and camera crews in patient areas. Federal laboratories, depending on the nature of their research, may or may not allow direct contact and require PIO involvement.

Media policies may also differ according to the topic or laboratory, and some are so absolute that you violate them at your peril. For example, a Duke neurobiologist once allowed a photographer for a national magazine into his laboratory in the vivarium, where animal research is conducted. The neurobiologist had wrongly assumed that the university's generally liberal media policy applied to animal research. He was blithely unaware that Duke prohibits media cameras in the vivarium and severely restricts journalist access. The violation was particularly perilous because he was doing invasive experiments on macaques, and he allowed the photographer to take shots of the animals strapped in restraining chairs. Fortunately, the photographer sympathized with the scientist's plight and, although not obligated to do so, agreed to crop the images submitted to the magazine so that they did not show the restraints. Although the researcher was reprimanded for the oversight, the matter ended there.

One invariant media policy is that institutions do not allow researchers to issue their own news releases. Rather, they are asked to work through their PIOs on formal institutional releases to the media. This policy is not intended to restrict academic freedom in the case of universities. Rather, it ensures that news releases are professionally done and disseminated and that the researcher is protected against his or her own inexperience.

However, the requirement for institutional news releases may not preclude your cultivating media contacts, sending journalists informal e-mails about your work, and posting explanatory material on your laboratory Web site. If you have any doubts about institutional media policies, ask your PIO to clarify them.

Internal Communications

Your institution may have a formal or informal structure for spreading news of scientific articles internally. In either case, be sure that everybody who needs to know about your papers is informed—preferably well before their publication. These people may include your department chair, vice president or vice provost for research, in-house publication editors, and of course your news office. A savvy PIO will offer to help disseminate news of your papers internally, says research communication consultant Cathy Yarbrough. "It's a useful favor, because sometimes scientists feel awkward about advertising their new research findings. As communicators, however, we can sound the bells and whistles and jump up and down about the importance of the finding," she says.

Your institution may have some formal way of spreading news of your publications internally. For example, while at Johns Hopkins, research communicator Joanna Downer edited an internal newsletter containing research briefs on newly published papers. The twice-monthly newsletter pulled descriptions of papers from news releases or research abstracts. The newsletter was enormously popular, says Downer. "A lot of times the scientists were much more interested in having their research in the science newsletter for an internal audience than they were in having a news release written that would go outside the institution," she says. If your institution does not have such a newsletter, consider suggesting that it launch one.

"Branding" Policies

The requirement that you use the accepted "branding" of your department, center, or institution might seem no more than a niggling bureaucratic detail. But institutional leaders greatly value such branding, so serve your best interest by citing your affiliations correctly in your news releases, feature articles, and Web site. For example, at Duke, all medical center news releases had to be branded as being from "Duke University Medical Center," as distinct from "Duke University," even though the medical center is part of the university.

Sometimes such branding directives are emphatically dictated by the institution. Yarbrough recalled the branding directive at the Yerkes Primate Center: "I was ordered that every time the Yerkes Primate Center was mentioned in the press, it had to be the 'Yerkes Primate Center of Emory University.' I can't tell you how much begging I did, particularly with broadcast reporters. I screamed, pleaded, almost cried."

Your PIO can tell you about branding requirements regarding the proper names of centers, departments, laboratories, professorships, and so forth.

Learn Your Unit's Communication Policies

In creating your communications strategy, find out how your own school, department, or center disseminates information internally. Ideally, its Web site features such information as news releases on its researchers. For example, the Duke Department of Neurobiology site, *Neuro.Duke.edu*, effectively highlights its researchers and their work. However, that department is very much the exception. The most widely used intradepartmental communication vehicle remains the article-festooned laboratory door.

Ask how your unit reaches out to important professional audiences such as prospective students, and colleagues and administrators at other institutions. How does it broadcast news releases and announcements of awards and professional appointments to those audiences? For example, some schools announce major appointments and awards by sending out handsome color postcards to the profession—including alumni, funding agency program managers, industry leaders, and other institution's administrators. Duke's Deborah Hill finds such announcements professionally valuable: "We are trying to chum the waters, so that eventually these faculty might be considered for fellowships in professional societies and other honors. It is nice to have these announcements as an additional communications tool besides news releases. It gives us a chance to directly reach editors and our faculty's peers, to highlight their work and to bring people back to our Web site."

Also, find out how or whether news about your research finds its way up the organizational chart. Your unit might produce reports of research advances, honors, and awards that go to senior administrators. Quite often, however, they learn such news right along with the rest of the organization, through news releases and in-house newsletters. Importantly, find out how your department notifies your news office about impending publications. Almost invariably it will be up to you to let your PIO know about your scientific papers.

Your unit can do a great service to you and your colleagues by notifying the news office of accepted papers. By far the best such notification system I encountered was that of the Caltech astronomy department. As part of the internal scientific review process for draft papers, their publication committee also rated each paper on scientific significance and newsworthiness and passed those ratings on to me as director of the Caltech News Bureau. These ratings told me which papers were scientifically important enough to do news releases on, including those I would have otherwise missed because they had arcane titles and abstracts. The result was that Caltech astronomy received the attention it deserved for its discoveries.

Besides understanding your unit's formal policies, also understand its informal practices and its culture of communication. Some centers or departments may encourage reaching out to lay audiences, while others may have a more insular culture. More likely, your unit has not even thought much about the place of such communications in its mission—even though its leadership is pleased when there is a media story about its work.

Finally, find out whether tenure or promotion decisions are influenced by your efforts at lay-level communications. Gleaning this information can be a complex and difficult task. In some units, for example, working with the media, local schools, or other audiences is counted as a plus in such decisions, while at others such activities are considered a distraction from research and publication.

Tell Your Whole Research Story

Your communication strategy should aim at giving a comprehensive, coherent view of your research, including how each project fits into that view. Do not leave any pieces out, even though they might not seem important. For example, a paper on a relatively modest technical advance might not seem worthy of a news release. But neglecting to do a story on that advance may leave a gap in the overall perspective on your research. And as chapter 9 explains, in the age of the Web even releases on relatively minor research findings will reach a broad audience—including researchers who can benefit your work.

Also, even a finding that you do not consider a major scientific achievement might prove very interesting to the public. Such a popular story might lure key audiences to an interest in your work. So, do not hesitate to communicate such findings. For example, Elizabeth Luciano, then a PIO at the University of Massachusetts Amherst, learned that engineer David Schmidt had used a mathematical modeling technique to explain why shower curtains billow inward when the shower is on. Luciano issued a news release over the objections of a dean, who declared the finding frivolous. However, the story got major media play, appearing in the *New York Times*, the *Boston Globe*, *Scientific American*, and the *London Sunday Times*. The story served to highlight the software Schmidt used in his analysis, to educate the public on the Bernoulli effect, and to present the university as a place where people enjoy their work.

Fit into Your Institution's Mission ... or *Not*

Your communications strategy should also take into account how you fit into your institution's research mission. You may be part of a precise strategy or a more loosely defined one, depending on the institution. For example, universities hire researchers as part of an overall academic strategy and usually give them freedom to define their research and to communicate it. But corporations and federal laboratories define more targeted missions for their scientists and engineers, and their work and communications must fit that mission closely.

"At a national lab, you are so much more closely tied to politics, and the mission of the day, and whims of the administration," says Hill, who worked at the Idaho National Laboratory. "And that makes it difficult for people to establish their careers, because it is not just their publishing record that counts. And they cannot just shift their research on a dime as the whim of the day changes." Thus, a communications-savvy researcher will recognize when he or she might not be part of the current central mission, says Hill. The researcher will need a deft communications strategy to position himself or herself to survive until that mission swings back toward his or her work. A good PIO can help you think strategically, beyond such passing whims, notes Hill: "You have to look at the strengths in your organization and think 'big picture' about how communicating your work contributes to maintaining that overall identity," she advises.

For one thing, a strategically thinking PIO can help you maintain the flow of communications during such temporary disconnects between your work and your institution's mission. Catherine Foster, former Argonne National Laboratory media relations manager, describes how she helped to preserve research projects against changing political winds: "Clearly, if you have a politically unpopular program your funding sources don't want you to talk about it," she says. "From that point of view you kind of go underground. But you don't want to lose the momentum, so we would work with trade press to continue to quietly promote such work with those media." And, says Foster, even though the laboratory's management would not authorize formal news releases about research projects that were "off-mission," she nevertheless found a way to spread the word about them: "When such people were presenting at meetings of the American Chemical Society or the Materials Society, we would work the phones before and after, telling media that 'so-and-so is doing interesting work and would be available to talk to you.'" Such communications proved ultimately valuable for the laboratory, says Foster. For example, the laboratory's Integral Fast Reactor program was shut down during the Clinton administration but became a major priority of the Bush administration. To help the program remain viable during its political "winter," the laboratory quietly continued to fund the nucleus of the program and Foster continued sub rosa communications about its achievements.

There also will be times when you believe your research findings need to reach an important audience, and your institution's PIO is just not interested in publicizing it. If so, the best route is to work with your PIO to figure out a compromise that will allow you to reach those audiences yet still remain within the institution's policies.

What Does Your Funding Agency Expect?

If you have NASA, USGS, DOE, NSF, NIH, or other government funding, your principal contact, of course, will be your program officer. You will want to understand what his or her preferences are in terms of communication. Your program officer may not necessarily be in touch with the agency's PIOs, though, so you should also understand the communications machinery within the agency. The special online section on working with PIOs at *ExplainingResearch.com* covers in more detail how the NSF and NIH communications offices work, and how they use news on your work and your findings. Such agencies deem such information invaluable in advocating their budgets with Congress and reaching other important audiences.

Become an Expert Resource

Your strategy should also include making yourself a useful expert resource, both internally and externally, beyond your own work. For example, as discussed in more detail in chapter 22 on meeting journalists' needs, you can offer to be a "mole" in your department—a source of tips for your PIO on interesting research in the department.

For the media, you can be included in expert lists maintained by your news service, your professional association, and other groups. For example, the Society for Neuroscience maintains a media directory that includes neuroscientists willing to talk to the media. Also, the AAAS operates Science Talk, an extensive listing of experts available to advise journalists and give talks at events. The AAAS research news service EurekAlert! also separately hosts an experts database of researchers willing to offer background and opinions to journalists on research findings. Your PIO can submit your name to that database. The Science Media Centre in London offers journalists expert scientists to comment on stories, as well as media briefings, workshops, and discussions for journalists, scientists, and PIOs.

The ProfNet service is another major source of experts, including researchers. It connects journalists and others with expert sources through queries posted by journalists or by a search of the ProfNet experts database. If your institution is a ProfNet member, you can be listed as an expert in its database.

You may worry that inclusion in such lists will bring an overwhelming flood of calls from journalists looking for help and quotes. Far more likely, you will receive a modest number of calls, perhaps none at all. However, you will still find it useful to be listed, since it puts you on the radar screen of the PIOs in your institution, as well as others whom you want to know about your work and expertise.

Seek to "Dominate the News Space"

Your purpose in communicating research should be to "dominate the news space," says science communicator Rick Borchelt. This objective is not as crassly publicity-seeking as it sounds. Borchelt does not mean spewing forth promotional news releases or publicizing yourself merely to satisfy your ego: "That elevates your brand, but it may dilute your impact," he warns.

> By "dominate the news space," I mean I want that if a *Washington Post* reporter is going to do a story about human genetics, that story is incomplete unless she talks to someone at the Genetics and Public Policy Center. To that end, when I was communications director there, I made sure that our materials on the Web were easily accessible to her and at a range of levels of sophistication, including easily understood issue briefs and more technical white papers linked to a technical piece in a journal. So, someone could choose the depth at which they went into the information.

Thus, your communications strategy should aim at portraying you as a thoughtful, articulate, substantive researcher whom media and other audiences naturally think of when they are listing leaders in your field.

The "news space" is not restricted to media. Your communications strategy should aim at presenting you as a reliable expert to all your important audiences. Thus, your strategy could include field trips for policy makers, workshops for teachers, and lunches with colleagues outside your discipline. For a good collection of such communications, see the Web site of the 2002 national conference "Communicating the Future: Best Practices for Communicating Science and Technology to the Public." Borchelt chaired the working group of PIOs, communications researchers, journalists, and scientists that compiled those best practices.

9

The Essential News Release

Just as the quark is the basic unit of particle physics, the news release is the basic unit of your research communications. And just as different flavors of quarks combine to form atoms, different types of news releases combine to form your high-energy "atomic" research strategy.

The Many Types of News Releases

Understanding the many news release content types is important because they can have very different uses and can target different audiences.

The Hard News Release

This is invariably justified by a "news peg," which is a reason to issue a news release that usually consists of publication of a scientific paper or a talk on your research findings. Other news pegs include submission of a formal report on a study or a discovery to your funding agency. The "hard news release" triggered by a news peg offers the most immediate payoff in terms of media coverage and public attention. Your institution or funding agency will issue most hard news releases with an embargo dictated by the journal or the meeting organizers.

You risk your credibility if you issue a news release and/or hold a news conference on findings not based on a legitimate publication or professional talk. The most notorious case of such an unwarranted hard news release was the 1989 announcement of the achievement of "cold fusion" by chemists Stanley Pons and Martin Fleischmann. The University of Utah issued a news release and held a news conference before the scientists had even published a scientific paper on the results. The university and the researchers were heavily criticized for the rush to publicity, not to mention that their work was generally discredited.

The Feature Release

This typically describes work in progress that has not yet yielded publishable conclusions. A feature release may be sent to the media and/or published in in-house newsletters and Web sites. It often includes anecdotes, first-person descriptions, conceptual background, and potential applications. It also leads off with a compelling beginning such as a human-interest angle aimed at attracting readers.

Feature releases are useful in calling attention to work that is inherently interesting but that has not yet yielded a result. They also provide background information on your work that you can send to key audiences. So, when you believe your work is far enough along to justify a story, encourage your institution to do a feature on your work. Be ready to provide the writer with the kinds of anecdotes and perspective on your work that make for an interesting feature.

The Backgrounder

This is much is like a feature release, but it is usually a more nuts-and-bolts description of the history and evolution of a piece of research. A backgrounder usually accompanies a hard news release as part of a media kit, discussed in chapter 10 on crafting releases, and can also be posted on your Web site. The backgrounder aims to give media and other audiences more detail on the work than should be included in the hard news release.

The Personal Profile

This is a feature article that concentrates on you as a person. It usually describes your personality and background, how you became a researcher, and your challenges and adventures as a scientist. The profile may also include your role in controversies, as well as comments from fellow researchers and others who know you.

Many researchers are uncomfortable with profile articles. After all, the scientific culture portrays research as an impersonal search for truth. You may not feel it is pertinent that you flunked freshman calculus or that your favorite hobby is building ships in bottles. Or, you may be embarrassed by such revelations as the fact that, while you are a brilliant theoretician, you are a klutz at the lab bench and your techs have pleaded that you stay clear of expensive instruments. However, the many uses of personal profiles should persuade you they are worth some minor discomfort. They portray you as an interesting person to many important audiences—for example, donors, prospective students, and lecture audiences—who want to know more about you. Such personalizing makes you a more accessible, even authoritative figure. Profiles also help students understand that researchers are people just like them, giving them confidence that they, too, could become a scientist or engineer.

The Q&A

This type of news article offers your opinions and explanations in your own words. The Q&A portrays you as an authoritative expert and/or institutional leader with something important to say. Typically, it is written for internal consumption, but media do sometimes use the Q&A format to bring a more personal feel to a topic. When your own public information officer (PIO) does a Q&A, you have considerable control over its content. You can suggest questions and review your edited answers to make sure they say exactly what you want to say. However, if the Q&A is done by an outside journalist, as with other news articles, you can offer to review the Q&A, but do not expect the journalist to comply.

The News Tip

This consists of a brief "nugget" that may either describe your research finding or highlight your expertise on a topic in the news. The news-topic-related tip aims to interest journalists in quoting you in their stories on the topic. Some organizations send out periodic collections of feature news tips to media. Often, such tips are backed up by feature articles that offer more in-depth information. Also, some journals, such as *Science* and *Nature*, prepare brief summaries of newsworthy papers in each issue that they send as a press packet to media.

The Media Alert

This release notifies journalists of a news conference or other event. It includes all the pertinent information about the event—place, time, participants, and so on. However, the alert does not include too much information about a news conference covering an embargoed scientific paper. Otherwise, journalists would have enough information to do a story, breaking the embargo. Embargoes are discussed in the "Working with PIOs" section at *ExplainingResearch.com.*

The Grant/Gift Announcement

This describes a new gift or grant, explaining the objectives of the research. Researchers and administrators quite often overestimate media interest in new gifts or grants—mainly because they are so delighted to receive the funding. However, "money does *not* talk" to most audiences. Even very large dollar amounts do not particularly interest readers and will likely warrant no more than a few lines in the local media. Chapter 10 on writing news releases includes how to write a grant announcement that is more likely to interest readers.

Fundraisers may push for a gift announcement to curry favor with the donor or foundation giving the gift. However, sucking up to donors is not a compelling reason to do a news release on a gift. Instead, such an announcement can be posted on an internal Web site and/or published in the in-house newspaper. Nor are news releases appropriate for promoting some fund-raising priority, such as attracting donations for a new laboratory. Rather than issuing such promotional releases, development officers should use the legitimate news about research achievements to illustrate the quality of the institution's work, which will be enhanced by that new lab.

The Award Announcement

This announcement of a prize or honor is another type of release that is seldom of interest to the external media, but which researchers and administrators love to see publicized. Such releases also aim at massaging the recipient's ego. Resist the temptation to ask your PIO to do a multitude of award releases. For all but the most important awards, more appropriate are notices in internal publication and on the institutional and departmental Web sites. The same brief notices can be e-mailed to the researchers' neighborhood weekly newspaper and/or Web site, which are more likely to feature the award.

The Many Uses of News Releases

News releases have a much broader utility than their traditional aim of attracting media coverage. In fact, these other purposes may outweigh even the prime objective of prompting news stories and can justify doing a news release even if the finding will not interest media. Beyond attracting coverage, news releases can serve as ammunition for your funding agency, background material a statement of record, a Web alert for fellow researchers, a record of stewardship of public funds, and many other purposes.

Ammunition for Your Funding Agency

NSF, NIH, and other agencies readily use outside news releases in their print and online publications. For example, the NIH's National Institute of General Medical Sciences searches EurekAlert! for news releases on NIGMS-funded research and puts them to multiple uses. They post releases on the NIGMS Web site as Research Briefs and use them in their e-newsletter *Biomedical Beat*. Also, NIH may use releases as the basis for stories in its *Research Matters* online magazine and in radio and video podcasts (vodcasts). See the online resources for this chapter at *ExplainingResearch.com* for links to these outlets.

Internal Communication

Releases can give your own administrators lay-friendly updates that help them understand your work. Administrators also may use releases as fodder for their own communications to their constituencies. For example at universities, writers for the president often draw on releases for their presidential communications. These include the president's report to trustees, the institution's annual report, and letters to legislators urging support for research funding.

Development officers also use news releases to court donors and potential corporate partners. For example, at Duke, fundraisers drew on news releases when courting for further donations a businessman who had given a major donation to a building in his name. As part of their cultivation, they sent him an annual report of the significant research advances made in "his" building—created by drawing on the substantive collection of news releases done on that research.

An Investment

Releases can serve as a communications "investment" in a promising young researcher. They provide useful background information, even though the papers they cover may have been rather technical and less than "newsworthy" to lay media. For example, I made such investments at Duke by writing features and releases about the early, basic studies of two young neurobiologists, Erich Jarvis and Michael Ehlers. In electing to do these releases, I had judged that these young scientists were going places in their careers. Their work was quite excellent, and it appeared in prominent journals.

Indeed, Jarvis was subsequently named a winner of the NSF's prestigious Alan T. Waterman award for promising young researchers. The news releases and feature articles I did helped the NSF gain national media attention for his work. And Ehlers and Jarvis were subsequently both named Howard Hughes Medical Institute investigators—which distinguished them as the cream of the scientific crop. Again, the releases and features gave audiences useful background on their work. They also gave the university administrators ready-made information about the scientists.

Such a communications investment can have a two-way impact, also teaching the researcher about lay-level communication. For example, Ohio State PIO Earle Holland recalls the impact of investing in coverage of the work of glaciologist Lonnie Thompson when the young scientist first arrived at the university. "I did the first story on him that had ever been done, and have done twenty-seven stories on him since then," recalls Holland. "During that period, he rose from being a rather ostracized researcher for his views on global warming to being a National Academy of Sciences member, and he has gotten just about every award in geological sciences you can get, including the National Medal of Science. He was Al Gore's chief adviser on *An Inconvenient Truth*."

"It was a true partnership, because as I was educated in his science, he became educated in communications," says Holland. "Also, when *Discover*, *Rolling Stone*, or *Time* highlighted him, they had a wealth of news releases to draw on as background on his findings." What's more, says Holland, such extensive external coverage of his work reflected back on Thompson's status and visibility within the university. His visibility made him an attractive candidate for honors and made it easier for him to advocate for funding and other resources.

Background Material

Releases serve as useful background material for subsequent stories. As NIGMS PIO Alisa Machalek says, "If I am writing a story about RNA interference, I will look up the releases we've done to see what we said. Or, if a journalist writing about RNAi calls and says 'I need some background fast; I have a story due in two hours,' the background is all written up already. It has already been cleared and is posted on our site."

A Statement of Record

A news release is also a public statement of record on a piece of research. Unlike a scientific paper, a news release constitutes an accessible lay-level account of your work. It can include your assessment of the implications of your work and your future plans. It also constitutes a public acknowledgment of credit to colleagues, which could protect you against charges that you are trying to hog the glory. In contrast, media stories never list all the research collaborators and may misrepresent their roles.

The news release is also a historical document. Over the decades, I have been privileged to write releases on some of the major developments in science, from the synthesis of the first gene at MIT in 1970 to the announcement of the first neural control of a robotic arm in 2003. In both those cases, and in many others, it was particularly important to provide an accurate, detailed lay-level explanation of the research, since that information would become part of the historical record.

A Web Alert for Fellow Researchers

A news release is also searchable information about your paper or talk. Scientific papers or meeting abstracts may not be picked up by search engines, but news releases are. In fact, news releases often appear in search engines right along with media stories on a research paper or presentation. So, news releases posted online make it more likely that such audiences as prospective patients, prospective corporate partners, other researchers, and potential donors will find out about your work.

Says Howard Hughes Medical Institute PIO Jim Keeley, "We have investigators requesting releases, because they recognize that science is a competitive business. And they recognize that an HHMI release represents a lay-language record of the work out there on the Web that helps establish that they've made a discovery."

A Record of Stewardship of Public Funds

If you receive public funding, you have ethical and even legal obligations to account for your use of that support and to disseminate your research findings to the widest audience. News releases are the most important way to meet those obligations.

Education for the Public

The occasion of a research discovery constitutes a "teachable moment," offering a prime chance to educate the public about science, engineering, and medicine. You have the public's attention because you have discovered something new.

And the news release—along with images, video, animations, and other materials—offers the public an accessible way to learn about your finding and the science behind it. For example, as discussed in the introduction, the NSF recognizes the educational value of such dissemination in the Broader Impacts Review Criterion by which it judges grant applications. In its judging, among the questions the NSF asks is "Will the results be disseminated broadly to enhance scientific and technological understanding?"

In such dissemination, NSF explicitly recommends that its grant recipients "publish in diverse media (e.g., non-technical literature, and websites, CD-ROMs, press kits) to reach broad audiences." The NIH includes similar language in its grant contracts requiring recipients to make efforts to inform the public of their work.

Family News

As mentioned previously, news releases can contribute to family harmony by helping show why researchers may spend so much time closeted in their laboratories. Vanderbilt PIO David Salisbury recalls even aiding spousal communication with a release. "I did a release on this senior scientist in his 60s, who was studying lightning," recalls Salisbury. "When I saw him some time after the release, he said 'I showed the story to my wife, and she said it is the first time she understood what I did.' They had been living together for 40 years, and obviously he hadn't been working on his communication skills," quips Salisbury.

10

Craft Releases That Tell Your Research Story

An effective news release adheres to editorial rules that writers have found effective over long experience. If your releases observe these rules, the media and other audiences will perceive your work as more credible and significant. Here are those rules of an effective news release:

Make the Header Informative

The header of a news release contains the housekeeping information necessary for a journalist to write a story:

- The institution's name, address, and main Web site URL.
- The PIO's name and contact information.
- The principal researchers' contact information if institutional policy allows direct contact between researchers and the media.
- Embargo information—the date and time a story on the paper may be published or posted. Embargoes, as discussed further in the online "Working with PIOs" section, are usually only for papers being published or delivered at a scientific meeting. Otherwise, the release includes the notation "For immediate release" and the distribution date.

- Links to publication-quality images, video, audio, graphics, and/or animation.
- An editor's note, if needed, covering special information such as the researcher's availability.

Write a Clear, Compelling Headline

The headline is your first chance to lure audiences to read your news release, so the headline needs to be a clear, specific statement of the discovery that engages audiences. Preferably it uses an active verb. Vague, general headlines are less effective. For example, which headline do you think would be more likely to interest journalists and readers?

> Duke Medical Center Researchers Develop Experimental
> Brain-Machine Interface

or

> Monkeys Consciously Control a Robot Arm Using Only Brain Signals

The second is more compelling because it states more concretely and intriguingly what the achievement was.

However, many news releases do not have the natural draw of monkeys and robot arms, perhaps covering more technical advances. In such a case, an effective headline might instead highlight the overall significance of the work. So, rather than the technical

> Discovery Reveals Mechanism of Dendritic Spine Function

a better headline for the release would be

> The Calculating Brain: New Study Suggests That Neurons Are Built to
> Perform Simple Arithmetic

If a piece of work might be misconstrued as frivolous, a good headline can protect the researcher. It can emphasize the finding's application, helping ensure that the application is highlighted rather than the seemingly frivolous aspect. For example, recall the study cited in the introduction by neurobiologist Michael Platt, which found that monkeys would forgo juice rewards to see images of female hindquarters. Anticipating that the work might be portrayed only as "Monkeys Like Porn," the news release headline emphasized the study's application:

> Monkey "Pay-Per-View" Study Could Aid Understanding of Autism

A headline that uses a metaphor or vivid phrase can attract readers to even the most basic research discovery. For example, here are headlines about some very basic discoveries in neurobiology:

"Reset Switch" for Brain Cells Discovered
Protein Facilitates "Hard-Wiring" of Brain Circuitry
Researchers Discover "Doorways" into Brain Cells
How the Neuron Sprouts Its Branches

Good headlines put key phrases up front, to attract readers. For example, the following examples fail to put the key phrase first:

Duke Medical Center Researchers Develop Experimental Brain-Machine Interface
Discovery Reveals Basis of Dendritic Branching in Neurons

The first headline starts with the self-serving phrase "Duke Medical Center Researchers Develop..." and the second is too technical and uses the flat phrase "Discovery Reveals Basis of..." As mentioned previously, better versions would be

Monkeys Consciously Control a Robot Arm Using Only Brain Signals

or

How the Neuron Sprouts Its Branches

Writers are often tempted to put the institution's name first in headlines, under the mistaken belief that it enhances the institution's name recognition. However, that benefit is more than offset by the resulting loss of "punch" in the headline. Emphasizing the institution also tends to give the release the air of a publicity piece, rather than the news story it should be perceived as.

Eliminating unnecessary words improves headlines. So, for example, even the good headline

The Calculating Brain: New Study Suggests That Neurons Are Built to Perform Simple Arithmetic

Could be improved by tightening it to

The Calculating Brain: Study Suggests Neurons Perform Simple Arithmetic

Sometimes, however, brevity is not the best strategy for headlines. The science behind some releases may require a longer headline. For example, the headline

Monkeys Consciously Control a Robot Arm Using Only Brain Signals

was not the one that actually appeared on the news release. The actual headline was

Monkeys Consciously Control a Robot Arm Using Only Brain Signals;
Appear to "Assimilate" Arm As If It Were Their Own

The longer headline was necessary because the neurobiologist Miguel Nicolelis believed the headline should emphasize that the monkeys appeared to adapt their brain circuitry to control the robot arm as if it were a third natural arm. Although the fact that the monkeys assimilate the robot arm might seem to be only a technical detail, it has turned out to be fundamental to understanding of how the brain learns, as well as how easily humans might adapt to external "neurorobots."

The headlines above are also strong because they tell the reader what the story is about. Unfortunately, all too many headlines do not. Rather, they seek to be clever at the expense of being informative, for example,

Monkey See Robot, Monkey Do Robot

Such cute-but-uninformative headlines are not effective. Their aim is admirable: to attract readers using clever verbiage. However, they fail to serve the primary purpose of a headline: to inform. They thus reduce the release's credibility and try the patience of readers who want to know immediately what the story is about.

It is possible, however, to have fun with headlines and still be informative. So when a clever headline comes to mind that is also informative, by all means use it. For example, for a release on a new genetic technique dubbed "P[acman]," I wrote the headline

P[acman] Permits Precise Placement of Prodigious DNA

For a release about a discovery that infants are mathematically adept, I wrote the pithy

Baby Got Math

And for a release about a fungus that launches its spores at stunning velocities,

Corn Fungus Is Nature's Master Blaster

Avoid using a headline that suggests an immediate clinical application for a basic finding. Such headlines overpromise. In his book *Science, Money, and*

Politics, Daniel Greenberg dubs such overpromising "may journalism." Among the dubious headlines Greenberg cites:

- Worm Gene May Offer Key to Aging Process (*New York Times*, May 13, 1999)
- Gene May Promise New Route to Potent Vaccines (*Science*, May 7, 1999)
- Knockout Mouse May One Day Lead to Major Understanding of Human Kidney Disorder (NIH press release, August 30, 1999)

Greenberg asserts that "even with the de rigueur cautionary qualifiers, the excited formulaic reports of wondrous medical breakthroughs sometimes run so far beyond clinical reality that confessional correctives become necessary."

For grant announcement releases, writers are frequently tempted to highlight the dollar amount in the headline and "lede" (see next section for a definition). However, people are not interested in dollar amounts. To make a grant announcement more interesting, the writer can highlight the research to be carried out with the funding. So rather than the headline

> Acme University Researchers Receive a Gazillion Dollars to Study Soap Scum

a better headline would be

> Soap Scum Study Aims to Rid World of Slimy Scourge

The release would concentrate on the remarkable new soap scum analytical techniques researchers will use and the ultimate objective of the work. Only in the third or so paragraph would the release reveal that the study is funded by a new gazillion-dollar grant from the National Institute of Schmutz.

Grab with the Lede

The "lede" of a news release comprises the first sentences that tell the reader what the release is about. (A historical note: the spelling of *lede* was meant to distinguish it from the spelling *lead* and harkens back to the era when newspapers were set with hot lead type. The space between lines of type was called leading, and journalists used *lede* so as not to confuse the typesetter.)

A good lede is succinct and informative, usually of the general form "Researchers have discovered X about Y. The significance of the finding is Z." Academics have trouble crafting good ledes because they tend to "back into" the story. Their training in writing scientific papers compels them to start with background on a subject. So, an academic's lede might start out. "The study of Y has a long history, beginning with the Middle Ages…"

Effective ledes are not "clever"—seeking to engage readers without stating immediately what the news is. For example, while a concise lede might read

> A new genetically engineered mouse shows many of the same symptoms of schizophrenia as humans with the disorder. Researchers who developed the mouse believe it offers a powerful new pathway to exploring the causes of a disease that ranks among the most prevalent causes of disability worldwide.

A misguided "clever" lede might be

> The mouse cringes in the corner of its cage, refusing to nuzzle its cage mates as do its brethren. And it builds messy nests, unlike the tidy clumps of cotton that its fellow mice construct. Not just an antisocial outcast, this mouse has a genetic defect that makes it mimic schizophrenia.

Such a lede would be more appropriate in a feature release, because it tells the reader to expect a general explanation of a piece of research and not a concise summary of a discovery. "Clever" ledes, or those that back into a story, frustrate readers and reduce a release's credibility. In their daily lives, readers are bombarded with information, and they want a release to tell them efficiently what the story is about.

As with headlines, a lede that puts the institution's name up front is also a bad idea because it tends to lose readers. Basically, readers do not care *who* did the work, only *what* the news is. So, those ledes serve the institution best that first lure readers with the news. Only then does the release disclose the institution. The lede's greater readability means that the institution receives more attention than if its name was up front in the lede.

Place the Nut Graf High

Readers also want to know immediately why they should care about a story—dubbed the "nut graf" by journalists. The term is a contraction of the expression "nutshell paragraph." In the general form of the lede above, the nut graf is the sentence "The significance of the finding is Z."

While the nut graf must be clear and lay-friendly, it should only make claims with which you are comfortable. You need not speculate that your results may cure cancer or reveal the meaning of life—unless, of course, you truly believe deep in your heart that they will. Also, you need not justify every research finding as having a practical application, if it does not. A release on a basic finding need only state in the nut graf that it helps answer a significant basic question. If you clearly and intriguingly explain that basic question, readers will be interested.

Place the News Peg High

Besides wanting to know *why* they should care about a news story, readers want to know why *now*. Journalists base their decision about doing a news story on this "news peg"—the event that sparked the release, such as the publication of a paper or delivery of a talk. So a good release emphasizes the news peg, including the name and date of the journal or scientific meeting and the names of the principal authors. The year should be included in the date, since the release may persist on the Internet for many years, without an attached header that might give the date. There is no need to include the formal title of a paper in the news release.

Never be vague about a news peg. Never say a piece of work was published "recently." "If a press release says work was done 'recently,' and then we find out that it was published three months ago, we will have wasted our time working on the story, and it will be of no use now," asserts Julie Miller, who has edited *BioScience* and *Science News*. "However, if it explicitly said the work was done three months ago, we might hold it to use in a feature."

Posters offer news pegs that are just as legitimate as published papers and talks, so consider doing releases on your posters. For example, says former American Geophysical Union public information officer (PIO) Harvey Leifert, "At AGU, the poster presentations are not the also-ran work that wasn't good enough to make an oral presentation. We just don't have enough room and enough time during our meeting to have all the research be presented orally, so many of them become poster presentations."

Use an Inverted Pyramid Organization

News releases are traditionally organized as an "inverted pyramid." That is, they place the most important information first, place details of the story farther down, and place the least important information, such as background, last. This style originated in newspapers because it enables editors to cut a story from the bottom up without losing important information. Just as with writing ledes, researchers may find it difficult to switch their editorial gears to the inverted pyramid organization for releases. Their training in writing scientific papers makes them prefer presenting background first and working toward conclusions.

Make Explanations Concise

Researchers also have trouble concisely explaining their research in news releases. They tend to want to include technical detail that will turn off lay readers, including journalists. As with your talks, your news releases should summarize accurately and succinctly, skipping over unnecessary technical detail in order to clearly explain the basic concepts of your work. True, you might have spent your entire professional life tracing that biological pathway or creating that intricate theory of stress fractures in alloys. And you might feel that failing to include the names of every enzyme in that pathway or to list the myriad factors contributing to stress fractures somehow diminishes your work. However, the real failing would be to let unnecessary detail spoil the chance to engage a broader audience in the concepts you have worked so hard to develop.

Include Caveats about Your Findings

News releases should include important caveats high up. These are typically cautions to prevent readers from incorrectly assessing your work. For example, did your clinical trial include only a limited number of participants? Was it a phase 1 clinical trial that aimed only to test a treatment's efficacy? What are the research and clinical challenges to be met before a basic finding leads to a treatment? Far too many medical stories fail to emphasize such tentativeness, misleading readers and giving disease sufferers false hope.

Caveats are especially important because your news release will appear online alongside media stories in search engine listings. The media stories may well include those caveats, and the release will suffer by comparison. And the more clearly you can explain those caveats, the more credible will be your news release. What's more, revealing the limits of your experiments and the remaining unknowns engages readers, who love a good mystery. However, in observing the inverted pyramid organization, you can place less important caveats lower in the release.

Offer Broader Perspective

Putting your work in perspective engages readers and lends credibility to the release. How is your finding unique? Does it confirm or refute previous findings by other researchers? What previous work of yours led to the new study? Did other researchers' discoveries inspire your experiments? Were there particular surprises in your findings? Journalist/author Keay Davidson points out the value of such perspective to your credibility:

> One of the most important things a scientist can do is to swallow their pride and point out in the press release the larger context of their research....A piece of research gains credibility if the scientist says right from the start that this follows on work done by scientists X, Y, or Z. And even better, if they can give references to the papers. That spares me the embarrassment of going to the editor and excitedly showing him a story and later discovering that it isn't novel research.

Also, if your research has policy implications, be accurate and frank about those implications. Your credibility is at stake. *New York Times* environmental reporter Andrew Revkin reflects the attitude of many national reporters when he says, "The closer a paper gets to being policy-relevant, the more apt it is to have been misportrayed by the journal in its summary or in the press release of the university. And when you really push in hard, two-thirds of them go away right away because it is not a story."

Finally, a bit of relevant personal history about the research does interest readers, including journalists. Did you decide to do the experiments even though you had doubts that they would work? Did a persistent graduate student persuade you? Are there interesting anecdotes that illustrate the problems you faced?

Properly Credit Participants

Of course, the release should properly credit a paper's authors, but there are limits. If a paper covered in a release has only a few authors, it is feasible to include their names high up in the text without interrupting the flow and losing readers. However, a release on a paper with dozens of authors typically does not list all of them, since they are available in the paper itself. If politics dictates listing all the authors, however, put the list at the end. In some cases, a long list of authors can actually enhance a release's credibility. One example is a 2004 release on a paper by a consortium of 29 neuroscientists—led by Duke neurobiologist Erich Jarvis—who had developed a revolutionary new nomenclature for the structures of the bird brain. Each coauthor was a prominent scientist, and listing all the authors made the story more compelling by emphasizing that the nomenclature was developed by such an illustrious group.

If there are co-lead authors who contributed equally to a piece of work, it is a good idea to indicate that in the release. Also, it may be important to make clear what components of the work were done by each author or laboratory.

Journalists will find this information useful in deciding whom to interview about aspects of the work. In any case, avoid making it seem that you led a piece of work when you did not. Such an unwarranted claim can lead to public embarrassment.

To be safe, circulate the draft release to all the authors to make sure they are satisfied with their listing, or understand why they are not listed by name if they are one of a large cadre of authors.

Credit Funding Sources

Always acknowledge funding sources, whether government or private. You may even contact your funding sources to find out how they would like to be listed. They may have specific preferences. For example, while some NIH institutes prefer that they be cited by name, others prefer an overall citation to NIH. Foundations may also have distinct preferences. For example, Howard Hughes Medical Institute prefers that their investigators not list HHMI as a funding source, because the institute fully supports its investigators as employees of the institute and not merely as a granting agency.

Crediting your funding sources in the release helps the PIOs in the funding agency identify your release, for example, on EurekAlert!, so it can use the release for its own purposes. What's more, contacting your funding source may well lead to broader attention for your release. For example, the NSF and NIH highlight on their Web sites and in publications news of research findings by scientists they support. For more tips on working with agency PIOs, see the special section on working with PIOs at *ExplainingResearch.com.*

Make Titles and Affiliations Unobtrusive

While you should list people's titles and affiliations, try not to let this branding interfere with the flow of the release. For example, see how this huge clot of titles interrupts the story:

> "This was a really amazing discovery that will no doubt win us a big-money prize," said Dr. Nelson Haff, who is the Richie Rich Professor of Astronomy and Theoretical Astrophysics and Director of the Center for Really Big Astronomical Phenomena in the Department of Starstuff Studies at the University of Southern North Nevada. "We will follow up this discovery once we have a bigger telescope with more flashing lights."

Neither readers nor journalists care about such titles. Rather, on first reference, identify people only briefly—for example, "said astronomer Nelson Haff." Relegate lengthy titles and affiliations to a later paragraph. Also consider reserving the title Dr. only for MDs, and cite PhDs only by name, without the title.

Offer "Real" Quotes

Pithy, vivid quotes make a release more memorable and interesting. Also, quotes can credibly convey subjective information about a finding that may not be appropriate for the explanatory text. For example, such subjective statements as

> The researchers were surprised at their discovery that the quasar was the brilliance of a million suns.

need to be attributed. One way is to simply add a "said" to back up the information:

> The researchers *said they* were surprised at their discovery that the quasar was the brilliance of a million suns.

However, far better is to quote the researcher, to make the information more memorable and interesting:

> "We were really stunned when our analysis showed that this little dot of light we thought was a star was an immense quasar that outshone a million suns," said Haff.

A slightly dirty secret about quotes: writers of news releases and even media articles may massage quotes to make them clearer or more dramatic. So, when a quote in a draft release is not quite what you wanted to say, by all means ask for changes to improve it. PIOs sometimes even make up provisional quotes, which you can rewrite or delete as you wish.

Above all, remember that quotes need to sound like an utterance somebody really uttered, rather than a dry scientific statement. Researchers are often uncomfortable with colloquial quotes, preferring to "bland" them by removing personal or dramatic content or phrases during editing. And, they tend to clutter quotes with technical language or long sentences. Bland, cluttered quotes are less memorable and engaging, and they reduce readers' interest in your research.

A good writer also knows not to "step on a quote," prefacing it with information that reduces its impact, for example,

> The researchers said they were surprised at their findings. "We were really stunned when our analysis showed that this little dot of light we thought was a star was an immense quasar that outshone a million suns," said Haff.

Rather, a good writer will set up a quote, for example,

> Haff recalled the moment at the end of a fruitless observing run when he and his colleagues first obtained their results.
>
> "We were really stunned when our analysis showed that this little dot of light we thought was a star was an immense quasar that outshone a million suns," he said.

Writers may also use indirect quotes to convey subjective information, when there is no memorable quote:

> With Gleevec, the thousands of people a year in the U.S. who contract chronic myeloid leukemia now have a much better prospect for long-term survival, said Druker.

However, lifting quotes from scientific papers is a poor practice, because people do not talk in scientese.

Finally, quotes should advance the story of the research, not the political agenda of an administrator—whether at the institution or a funding agency. So, there is usually no reason for a release to include a quote simply as hailing the work as significant from a vice president, program manager, or other administrator. "Pat-on-the-back, me-too quotes make the quotee feel good but that's not my mission, nor is it the mission of my office," says Holland. "Specifically that mission is to reinforce and enhance the reputation of OSU research as world-class. Including something in a release that is obviously self-serving and useless to the news media is self-defeating," he says.

However, it is perfectly legitimate to quote an outside expert, even an administrator, who actually explains the significance of a piece of work, for example,

> "This discovery, we believe, is highly important, because it establishes a new pathway for understanding the genetic malfunctions that lead to pancreatic cancer," said John Doe, director of the Office of Cancer Genomics of the National Cancer Institute. "Few researchers expected that this particular gene played a role in this cancer, much less what appears to be a causative role."

Avoid Hype Words

A kiss of death for the credibility of a research news release is the use of subjective hype words such as "breakthrough," "pioneering," "leading expert," or "major discovery." To convey a discovery's importance, simply let the facts speak for themselves. For example, if a piece of research is a true breakthrough, the release need only state that the finding represents the first time that such a discovery has been made. As Davidson warns, "Don't ever oversell, because it only takes one case of overselling to lose my confidence. It will hurt not just the scientist, but it will hurt the scientist's institution."

Attribute Subjective Statements

More generally, attribute subjective statements and statistics to an authority—either a researcher or a source other than the release's writer. Such attribution adds to the authority and credibility of the release. For example,

> About nine out of ten people who contract the disease would normally succumb to it within five years, according to Doe.

Be Reader-Friendly with Technical Terms

Some technical terms are, of course, necessary to fully explain a piece of research, but use them judiciously. Include only those terms necessary to understand your work, and ruthlessly weed out the rest. For example, does the reader really need to know the names of the many enzymes in a pathway or the names of all the components of the new superconducting alloy you have invented?

Define technical terms upon first use, and place the definition right after the first use. Also, when you use the full name for an acronym, put the acronym in parentheses immediately after it—for example, "prostate specific antigen" (PSA)—but only if you will use the acronym elsewhere in the release.

Sometimes, you will need to introduce a technical term just to give something a name, but fully explaining the term would add unnecessary detail. In such a case, enclose the term in quotes to tell readers they only need the name and not the full background, for example, "a process known as 'adiabatic cooling.'"

To reinforce an acronym or technical term in a reader's mind, use it repeatedly throughout the release. For example, rather than referring vaguely to "the enzyme," use the specific name. And consider spreading the introduction of technical terms in the text, so that readers are not bombarded with a confusing fusillade of them in one paragraph.

Include Comparative Measures

If you tell readers comparatively how small, big, long, or short something is, they will have a better concrete grasp of research concepts. Familiar comparisons also enhance the release's interest. For example, science writers often compare microscopic objects with the width of a human hair—about 200 micrometers. For larger objects, there is the period at the end of a sentence, the circumference of Earth, and the distance from New York to Los Angeles (or some other recognizable landmark). For tiny volumes, a good comparative is that a nanoliter is roughly the volume of a snippet of hair that is as long as the hair's width.

Also, convert metric measures to English. A good site for doing this is the Megaconverter, referenced in the online resources for this chapter at *ExplainingResearch.com.*

Invent Vivid Analogies and Descriptions

The vivid analogy or descriptive name compellingly describes research concepts for lay readers and can have surprising benefits for your research. Such an analogy or description can engage lay-level audiences, from legislators to venture capitalists to administrators, who would not otherwise resonate with prosaic explanations of your work. What's more, such phrases may find their way into the scientific jargon. Thus, do not hesitate to invent such analogies or labels. A few examples—with apologies for making you wait this long after the teasing mention in chapter 1:

- **Artificial dog.** Cornell veterinarians invented a chamber for growing fleas, and we were trying to describe it in a news release. The chamber had many functional characteristics of a dog—a skin-like membrane through which the flea could bite to feed, a supply of warm blood as nutrient, and a clump of dog hair for nesting material. Given these dog-like properties, I suggested dubbing the invention the "artificial dog." The researchers were dubious about the seemingly frivolous name, but they went along. The name was a howling success (pun intended); the release garnered considerable publicity, and the scientists ended up patenting the invention as the Artificial Dog. And, the device greatly accelerated research into flea physiology and flea-control devices.
- **Cosmic blowtorch.** When Caltech astronomers discovered a "relativistic high-energy beam" of particles emanating from a quasar, I suggested naming it a "cosmic blowtorch." The name is now part of the scientific vernacular for the objects.

- **Anaconda receptor.** To vividly portray the structure of a "seven-membrane-spanning receptor," which winds itself in and out of the cell membrane, I dubbed it an "anaconda receptor." The researcher has subsequently used the metaphor in his own explanations.
- **Shotgun synapse.** When neurobiologists developed a new computer simulation of the synapse—the connection between nerve cells—they wanted to convey the explosive launch of neurotransmitters across the connection. I suggested terming the concept the "shotgun synapse," which was subsequently used in news stories about the simulation.
- **Jellyfish cells.** To convey how dendritic cells in the immune system reach out and attach to other cells, I described them in a magazine article as "like microscopic jellyfish extending tangles of delicate tendrils that entwine themselves about neighboring cells."

Stanford climatologist Stephen Schneider's explanation of the mechanism of global warming is an excellent example of a vivid analogy memorably explaining a complex concept. He portrays the stochastic nature of global warming by saying "climate is like a die: it has some hot faces, some wet faces, some dry faces, etc. I think our (in)action on global warming is loading the climatic die for more heat and intense drought and flood faces."

Make a Clear Conflict-of-Interest Statement

If the research involves any corporate partnerships, include an explicit statement about whether you or any other coauthors hold any financial interest in the company or have acted as a consultant or lecturer. Regardless of how minor such involvement has been, it is best to state it. Even a conflict-of-interest statement indicating no financial involvement shows that you have addressed the issue and lends credibility to the release.

Produce Compelling Visuals

Later chapters cover in detail how to produce compelling, informative photos, animations, and video. Such visuals should be an integral component of your news releases. In many cases, they will determine whether your news release is picked up by the media and whether readers will be attracted to read it.

For example, when he worked at the *Chronicle of Higher Education*, University of California–San Diego PIO Kim McDonald recalls that availability of images frequently determined whether the *Chronicle* used a story, since the images attracted readers to the science section. So as a science communicator at UCSD, McDonald makes sure news releases are rich in visuals. In fact, he sometimes sends out only photo news releases, which often get better use than releases, because they are so succinct, he says. Also he points out, even print publications will use video snippets on their Web sites.

Explain the Work Comprehensively

A news release should be a comprehensive explanation of a piece of research, rather than merely a short summary. Such depth is warranted because a news release serves a range of audiences, although some PIOs mistakenly advocate short news-nugget releases that serve only as media alerts. "PIOs can do a real service by crafting more comprehensive news releases," says AAAS public programs director Ginger Pinholster. While the AAAS and other journal publishers do issue short lay-level media summaries, she says, "our mandate is simply to convey the punch line of the paper as it was accepted, with no embellishment or interpretation. However, the institution can write a release that puts the work in full perspective."

Even if a news release is aimed only at the media, it still needs to be more than a short news nugget aimed at newspapers, which usually do short stories. It needs to explain the advance in enough depth to enable science media, such as *Scientific American* and *Science News*, to decide whether a research advance is worth covering in depth. What's more, since the news release is the public statement of record on a research advance, it should be comprehensive enough to effectively counter errors in media stories. It also serves as a more complete information source for important audiences such as other researchers and educators.

Duke research communicator Joanna Downer also points out that comprehensive news releases on clinical advances—especially including any caveats about a new treatment—serve an important medical purpose. "If you don't include the full details, you are doing a huge disservice to potential patients and to the physician or researcher. Patients could be given false hope, and the physicians are going to be inundated with inappropriate contacts," she says.

Adapt the Release for the Web

The unique medium of the Web imposes editorial requirements on news releases beyond those discussed above. For example, Web-friendly releases should follow the same layout guidelines for making text scannable as discussed in chapter 7 on designing your Web site. For example, instead of putting lists in paragraph form, make them into bullets. Web releases need particularly tight, informative headlines, subheads, page titles, and key words. Similarly, the text that appears in the Web page title bar should clearly explain the content. The page title is usually the release headline.

The text of a scannable Web release should consist of shorter sentences than you might otherwise create just for print reading. Also, this text should be chunked into shorter paragraphs, with one idea per paragraph. For easy scanning, paragraphs should be no longer than the width of the text column. And text columns should be narrow, typically a third to a half screen width. Text should not stretch all the way across the screen. As mentioned in chapter 7, an excellent source of tips on Web writing is the book *Hot Text: Web Writing That Works* and the corresponding Web site, *WebWritingThatWorks.com.*

Adapting your release for the Web enhances readability, as found by the eye-tracking study by Jakob Nielsen and Kara Pernice Coyne mentioned in chapter 7. They found that reformatting content for the Web—with bulleted items, subheads, and tighter writing—increased comprehension by 12 percent and increased reader satisfaction for online readers.

Ledes on Web releases must be tight because search engines tend to chop them off. So, for example, a lede that puts the name of the institution up front risks becoming uninformative when chopped by a search engine. Here is how the lede on such a release from Duke University was chopped on the Google News listing:

> Duke University Medical Center researchers have discovered that activation of a particular brain region predicts whether people…

In contrast, a Reuters news article on the same topic, when chopped, still yielded basic information on the story:

> Altruism, one of the most difficult human behaviors to define, can be detected in brain…

Instead of posting a complex release as a single Web document, consider whether you can organize it into an efficiently written "backbone" containing the basic information, with the less important information—background, technical explanations, bios, etc.—relegated to secondary pages with links from the main page.

Your Web releases also will be more involving and credible if you provide links to background information or definitions of key technical terms. These links can lead to content on your site, as well as be "outbound" links to authoritative outside sources. Do not worry that outbound links will take your audience away from your site. Users tend to return from such outside links to the originating site, Web use studies have shown. Generally, internal links open in the same window, while outbound links open a new window. A collection of such outside references is included in the online resources for chapter 7 on developing your Web site.

Your Web releases should also integrate high-quality news photos, animations, audio, and other multimedia—giving users the ability to download publication-quality images.

Finally, text news releases that are e-mailed should include a link to the Web version, which can give readers a richer resource, complete with multimedia and links to background information.

Create a Media Kit

A media kit is a collection of materials, which may be posted as a separate category on a Web site or produced on paper, that offers background on a research finding, project, or facility. It enables you to respond quickly and completely to the information needs of both media and your other audiences.

Among the possible components of a media kit:

- Current and past news releases and features on the research
- General description of the research in your laboratory
- A backgrounder on the specific research project, center, or facility
- A backgrounder on the institution
- Bios of the principal researchers
- Frequently asked questions about the research
- Major media clippings, or links to them in the case of an online kit
- A gallery of news images portraying your work
- URLs or links to multimedia, including audio, video, and animation
- General background information on a topic from credible sources, as print copies, URLs, or links

Of course, you or your PIO may have already produced many of these materials, which can be readily incorporated into the media kit. As you produce them, keep in mind their possible application to a media kit. For example, if your laboratory research description is highly technical, you may want to produce a lay-level version as well, for your Web site and for a Web or paper media kit.

11

Target Releases to Key Audiences

Your public information officer (PIO) will manage distribution of your news release to media. News offices usually maintain their own media lists and/or subscribe to services, such as Cision and Vocus, which maintain lists of national and international media, searchable by topic. The news office can also post the news release on research news Web sites that serve the media and the public. These include EurekAlert!, Newswise, and the European site Alpha Galileo. The office may also post the release on the general newswire Ascribe, which distributes a broad range of news from nonprofit organizations. Besides reaching the media, such posted releases also show up along with other media stories in such news aggregators as Google News and Yahoo! News.

Your PIO may also pitch the release to the sponsoring agency, for example, the NSF or NIH, to interest them in taking the lead on the release and/or highlighting it on their Web site. An NSF or NIH release gives your work the prestige of being highlighted by a major funding agency and can attract more media coverage of a research finding.

Remember Internal Media

Do not neglect in-house media such as campus newspapers and alumni magazines; they can be very useful. For example, university alumni are naturally interested in work from their alma mater and may advocate for your work in their own organizations. In fact, they may even turn out to be funding sources.

Deborah Hill, communications director for Duke's Pratt School of Engineering, recalls struggling to convince one reluctant researcher to cooperate on an article on his work for the Pratt alumni magazine. But the article paid off, says Hill. "When we sent the magazine out to the alumni, a guy who had graduated from here thirty years ago was so excited about what was going on that he cut a check for $50 K and said 'I want to establish a discretionary fund because I really want this to work. I want to help this in some way.'"

Target Trade Media

Similarly, emphasizes Hill, trade media stories can also have important impacts. "Everyone always thinks we should get into the biggest mass media outlet possible like the *Wall Street Journal*. But I have seen some of the biggest impact come from focusing on trade journals and niche magazines." When Hill worked at the Idaho National Laboratory, she achieved gratifying results from efforts to gain media attention for an engineer's work on a new metal coating:

> We got a tremendous amount of media coverage in trade journals. But it was really below the radar of the administrators of the lab until we started getting calls from all these businesses. And over a years' period of time, we pulled in a million dollars' worth of use licenses from companies who wanted to see whether they could use the coating in their product. Yet the administrators didn't feel like the communications campaign was effective.... They were only concerned that the story didn't get into the *Wall Street Journal*, the *New York Times*, or the *Los Angeles Times*.

Do Not Spam

Some PIOs mistakenly send a blizzard of inappropriate releases—on hard news, grants, awards, and so on—to all media, hoping that a release will prompt a story. Also, such spamming enables them to say the release went out widely. Since e-mail enables essentially free release distribution, such spamming is unfortunately very easy.

However, such an approach could lead journalists to overlook the really important stories. Editor Julie Miller cites releases she received from Wake Forest

University as an example of spamming. "Maybe it is a fine research university, but I get an e-mail from them almost certainly three times a week. And I can't believe they are doing that much good research. I just see Wake Forest on the return address, and I delete it. So if they have two good stories a year, I am not going to get them."

In contrast, selectively sending releases only to journalists likely to use them builds a reputation for producing quality information that is in your long-term interest as well as your institution's. The ultimate objective in selective distribution of quality releases is to "annoy" reporters, as former Associated Press reporter Lee Siegel once told me I annoyed him. "Your releases really 'bother' me because I know when I get one I'll have to open it, read it, and do a story," he told me.

News offices should try their best to ensure that journalists want their releases, for example, by querying journalists on their topic preferences and creating targeted mailing lists.

Do Not Flack Releases

All too many administrators and researchers believe wrongly that a PIO should aggressively "flack" a release to get the maximum media attention for it. This means indiscriminate phoning and e-mailing journalists to pitch a hard news release. Journalists generally find such pitching annoying, except when such phone or e-mail messages alert them to truly significant news they might otherwise miss. While such flackery might seem like a good tactic to sell a news release, it is a poor long-term strategy, reducing the credibility of your research and your institution.

In contrast, allowing a release to stand on its own, without phone calls and e-mails, motivates journalists to pay attention to each news release. They know that they cannot depend on the crutch of receiving a phone call from a PIO bugging them about each release.

Offer Advice on Distribution

Help ensure that your release reaches important audiences by supplying your PIO with the names of publications and people whom you would like to receive it. This list should include institutional administrators, your institution's foundation and corporate relations officers, funding agency program directors, and donors who might be interested in the research.

Also include fellow researchers and grant officers in your distribution. While they will probably read your paper or attend your talk themselves, they may find a release useful in explaining your work to people who do not have a scientific background. For example, your grant officer might want to send the news to an agency administrator who needs a lay-level explanation of the work.

Many news releases are "embargoed"—that is, journals may specify a date before which the media may not publish articles on the findings. That embargo date and time is included with the news release, and journalists are expected to observe that embargo. While you can talk to a journalist ahead of time, with the agreement that the journalist will observe the embargo, do not break the embargo by distributing an embargoed release to colleagues or other audiences. Only media should receive a release before the embargo. However, after the embargo is lifted, you can do your own personal distribution, for example, to family members. Also make sure the release is posted on your Web site as well as your department's. See the online *ExplainingResearch.com* section "Working with PIOs" for a detailed discussion of embargoes.

Follow News Release Etiquette

There are codes of etiquette regarding news releases from multiple institutions about the same piece of research:

- **If researchers from multiple institutions contribute equally to a piece of research, each institution can appropriately do a news release.** Each release can quote that institution's researcher. However, the release from the principal author's institution should be the one most widely distributed to the media. The other institutions can distribute their own versions to local media and post them on their Web sites.
- **If you are not the principal author of a paper, your release should make that clear.** It should list the lead author and institution first in the citation. Such collegiality and forthrightness will pay you back in greater credibility for you and your institution.
- **If the work underlying a paper was evenly divided among multiple institutions, it is statesmanlike to quote the senior author from each institution in any release on his or her component of the work.**
- **Some situations may warrant joint or simultaneous releases.** Such release distribution should be coordinated to avoid confusing the media with competing releases. The multiple releases should make it clear that other institutions are also issuing releases. However, there is no need to

discourage multiple releases. In fact, they may increase the likelihood that the media will use a story, says National Institute of General Medical Sciences (NIGMS) PIO Alisa Machalek:

> I have heard journalists say they use the "rule of three," that if three institutions involved in a discovery issue news releases, it is important enough for them to pay attention. It does make a paper look more important if three different institutions pay attention to it. So, if a finding is really significant, we don't inhibit ourselves from doing a release because of the chance that multiple releases will confuse a reporter. But in most cases when the institution is doing a release, we will just offer a quote about the significance of the research.

- **It is good etiquette to keep all the coauthors informed about the release.** It may not be necessary to circulate the draft to all of them, but they should know that a release is being done.

Advertise Your Clippings

Once media stories on your release appear, take maximum advantage of them. Ensure that your major audiences see them. Researchers and PIOs often mistakenly assume that, because a piece of research has been prominently featured in the media, everybody will know about the stories. A good example of such proactive dissemination is the process used by Rick Borchelt to remedy the lack of visibility for important media stories, when he was communications director at Oak Ridge National Laboratory. "Here was an institution focused on media stories as their goal. So, they would get this great story in the *New York Times*, and they would mention the story when visiting their congressman, and he'd ask 'What story?'"

Thus, Borchelt began sending major media stories, along with a letter from the director, to all important members of Congress, as well as program officers and directors of the DOE, which funds the laboratory. "The letter would say 'Thank you so much for giving us the opportunity to work on this exciting research that is reported in today's *New York Times*,'" says Borchelt. "And we would often append the news release, because it tended to have a better explanation of what we were doing; and also because we could acknowledge the funding agency, so they could see that they got credit. It was hellishly labor-intensive, but incredibly useful." Members of Congress would use the articles in their communications to colleagues, and even read them into the *Congressional Record*, says Borchelt.

Especially distribute trade journal news stories to important audiences such as foundations and potential corporate partners. Such stories are more likely to be missed, yet they may be more central to advancing your research than even stories in major national media.

Heed These Cautions!

Here are some key news release pitfalls that you can avoid if you know about them.

Be Meticulous with Statements on Data and Primacy

You were excruciatingly careful checking the data in your research paper; be just as careful with the lay-level interpretation of those data in your news release. Missing even a single qualifying phrase in a release can prove publicly embarrassing.

A good example was a news release by a Harvard-affiliated hospital on a study of complications from silicone breast implants. Don Gibbons—formerly communications officer at Harvard, and now with the California Institute for Regenerative Medicine—recalls the problem: "The researcher didn't look at the release carefully enough," he says. "The release said there was a negligible two percent increase in complications. But the increase was two percent *per year*, not over the life of the implant. The 'per year' wasn't in the release, and the error was pointed out in the *Wall Street Journal*."

Also, if the release claims some form of primacy—being the first, largest, most, and so forth—make very, very sure of your primacy, or else include a caveat that it is the first "as far as is known."

Contractually Protect Your Right to Publish and Publicize

If you are negotiating a corporate research contract as an academic researcher, ensure that you preserve your right not only to publish but also to freely publicize your work. The best way to preserve this right is to work closely with your contract office in the negotiations.

As a notorious example of such a failure, Gibbons cites the case of research on Synthroid, used to treat thyroid problems. In 1986, Knoll Pharmaceuticals, the company that manufactures Synthroid, contracted with researchers at the University of California, San Francisco (UCSF) to do a study comparing the effectiveness of Synthroid and other brands and generics of the drug.

The research found that the generics were as good as the name-brand drugs. However, unfortunately, the researchers' contract—negotiated without working with the university contract office—prevented them from publishing their data without Knoll's permission. So, under threat of a lawsuit, UCSF pressured the researchers to withdraw a paper already accepted by *JAMA* before it went to press. "The right-to-publish clause is basic in all contracts, and unless you make sure it is there, you will end up being as notorious as these researchers," warns Gibbons.

Beware of Exclusives

You or your PIO may be tempted to offer exclusivity on a release to a reporter at a major newspaper or magazine, in order to induce that reporter to cover the story. However, such exclusivity is illogical, inappropriate, and even unethical. Although you may make one reporter happy, you risk alienating all the others—which hurts your long-term credibility. What's more, the exclusivity will reduce overall media coverage of the story. AAAS's Pinholster cites a particularly egregious case. "I learned of plans by a public relations firm to post embargoed news releases to EurekAlert!, thereby establishing a specific embargo-release time for all registered reporters," she recalls. "Yet the firm wanted to give key media outlets permission to run the article early as part of an 'early advance exclusive' deal. It goes without saying that it is unfair and unethical in the extreme to tell all reporters to hold to an embargo, but then give one or two reporters a special deal—thus double-crossing everyone else and undermining the credibility and utility of embargo policies in general," she says.

In particular, veteran science writer Robert Cooke warns against what he calls the "*New York Times* syndrome":

> People will jump through hoops to get in the *New York Times*, offering them an exclusive story. In one case, the University of Pittsburgh medical school offered a *Times* reporter an exclusive on a genetic engineering study they were going to do on treating arthritis in a patient's knuckles. They did get in the *Times*, but it was inside the paper and didn't get a lot of attention. I was angry at them, because I worked for *Newsday*, a *Times* competitor. And after that, the PIOs at the university were also in trouble with the local paper, whose reporters wouldn't touch them after that.
>
> If they had sent out the release to everybody at the same time, treating them fairly, AP would have picked it up, and it would have gotten national coverage.

By far, the best route to publicity is to issue to *all* media a well-written release, along with professional-quality images, audio, video, and animations, as appropriate.

It is perfectly acceptable, however, to tell reporters at leading media about an important research finding in advance of other reporters, with the proviso that the reporters agree to honor any embargo. The objective is to give the reporters better access to scientists before other media, to give them more time to do interviews and prepare their story. However, such early access is not appropriate when a journal specifies a time—in *Science*'s case, the Monday before the embargo release date—before which researchers may not talk to the media about their papers.

It is also perfectly acceptable to pitch individual reporters on exclusive feature stories about your work. For example, you may be launching an expedition or starting a dramatic experiment about which a reporter could write a compelling feature article.

Sometimes a reporter may want to break an embargo because he or she learns the details of a research finding before the embargo date, or before you wish to talk publicly. The reporter may plan to immediately publish a story, regardless of whether you cooperate. In such cases, some horse-trading may be in order—offering the reporter early and complete access and a "one-cycle" jump on the story, in return for the reporter's cooperation in delaying their story. A one-cycle jump means that, for example, if a reporter works for a morning newspaper and the embargo is in the afternoon, the reporter agrees to publish the story and post it on the newspaper Web site that morning. In any case, notify both your and the journal's PIOs of such problems, and work with them on any solution.

There Is No Such Thing as an "Internal" News Release

Another basic mistake of researchers is to believe that news releases can somehow be restricted in their distribution. More than once a researcher has instructed me, "This should only be distributed internally. I don't want it to be public." Assume that any news release posted on the Web or even printed in an internal publication will be seen by the whole world, and manage it accordingly.

You Are at the Mercy of the News Day, the News Hole, the Reader's Roving Eye, and Fate

Editors at any media outlet unceremoniously toss your precious piece of research news into the story hopper along with the latest celebrity scandal, politician's malfeasance, or disaster. Thus, your media coverage is very much at the mercy of events and the limited-capacity "news hole" of print space or broadcast air time. Also, your breaking news ages quickly; editors empty the story hopper each day.

So, be prepared for the eventuality that deluges of other news can inundate your own story, and that not even the most expensive, sophisticated publicity effort can rescue your story. Science communicator Lynne Friedmann recalls a classic, and instructive, case of one of her clients: "A group planned a press conference to announce giving its entrepreneur/inventor of the year award, for the first time to a woman scientist/biotech CEO," she recalls. "To 'ensure' media coverage the woman's company paid sixty thousand dollars to one of the big PR agencies. As it happened, she received that award the day that [the Cuban refugee boy] Elian Gonzalez went back to Cuba." Buried by massive coverage of the boy's return to his father, says Friedmann, there was "not a word, headline, nothing came out in the media about the award, and the opportunity was gone."

If you are doing basic research, be realistic about the likelihood of broad publicity on your finding. As Duke research communicator Joanna Downer says, "Most scientists and some PIOs tend to think emotionally rather than logically about what the practical outcome is likely to be from a news release. If it is a basic science story, there are probably two dozen outlets, not hundreds, that are likely to carry the story. That really surprised me when I moved out of the lab to become a PIO."

Another reality is that even a widely publicized story will reach only some readers. Newspapers and news Web sites are vast compendia of information, and readers spend only limited time reading them. So, even the most compelling research news story may be missed by the huge majority of the public. Fortunately, the Web has an endless capacity for news and more permanency than a print newspaper. So you can at least be assured that your news will be available online to all who are interested. Also, since Web news is searchable, your information can be easily available for those who look for it.

12

Produce Effective Research Photography

Fortunately, much research is quite visual, so with some investment of time, effort, and money you can likely develop arresting images to tell your research story in a compelling way. You can also synergize your lay-level photography with your research photography, benefitting both. This chapter shows you how to make the most of the inherent visual nature of your work.

Even your technical images can be highly aesthetic while remaining accurate—as vividly illustrated by the work of MIT photographer Felice Frankel. She offers excellent tutorials and examples in her books *Envisioning Science: The Design and Craft of the Science Image*, and *On the Surface of Things: Images of the Extraordinary in Science*, with George Whitesides. Frankel emphasizes that visually communicating science not only benefits audiences, but also the researchers themselves. "Seeing science and making a visual representation of that science—that is, the process of thinking about how to visually represent it—clarifies the science for the person making the representation," she told an audience in a 2006 illustrated talk at the New York Academy of Sciences. A link to the talk is included in the online resource section.

Compelling scientific images also attract the attention of colleagues and give your work a professional image...literally. And, because striking images stimulate the visual center of the brain, they create a more visceral positive attitude toward their subject, in this case your work. And most practically, compelling images can make your research paper a candidate for a journal's cover.

Unless you are highly adept at photography, hire a professional even to shoot your research images. You might be surprised at how much better even the most prosaic instrumentation looks when shot by a professional. "It is one thing just to compose a standard technical shot; it's quite another to creatively compose it and light it in an interesting way to show detail and make the piece of equipment enjoyable to look at on a Web or print page," says Chris Hildreth, Duke's director of university photography. Certainly, for news photos always use professionals. They are adept at managing the subject, lighting, framing, and composition requirements to create a compelling news photo.

If you plan to do your own photography, consider taking a course and reading Frankel's books. Even if you will hire a professional photographer, her books will give you invaluable information. Also, study images in your field, and learn the production details of images you most admire. For example, the best in photomicrography is on display in the Nikon Small World Competition. Nikon also operates the MicroscopyU Web site, which contains tutorials, image galleries, and other information on producing quality photomicrography.

Control Your Images

Just as issuing a news release enables you to control the public information about your research, creating your own photography enables you to control the images. In contrast, if you do not produce your own news photos, the media will likely send their own photographers, and you will be at their mercy. "With downsizing at newspapers, there are fewer photographers on staff, shooting more assignments, so they have less time to spend on each assignment," warns Hildreth. "Thus, if you have a quality image produced by your own photographer, there will be no reason to dispatch a photographer. Your stock shot will be available and will be the one distributed to newspapers and magazines."

Hire a Good Photographer

Your institution might have a talented staff photographer, or you might have to find one yourself. If you need to hire your own photographer, Hildreth advises asking photo editors at publications whose images you like for recommendations. Specify that you are looking for a photographer accustomed to working in a laboratory environment. Review the candidate photographers' portfolios, interview them, discuss prices, explain what you want the images for, and select one who meets the criteria below. Whether you use a staff or outside photographer, you might also want technical as well as media shots. If so, make sure the photographer can produce both.

Once the photographer begins, you can detect the signs that you are working with a quality research photographer, according to Hildreth. For example, he says a good research photographer should

- **Ask good questions.** "A researcher should get good questions from the photographer about how he or she or the team did the research. Then the photographer will come up with ways to incorporate the lab instrumentation or perhaps bring in materials from field research to come up with creative images." You will usually discuss such questions during a location scouting session before the photo session, as discussed below.
- **Bring the right equipment.** "A photographer from a newspaper will generally walk in with a bag over his shoulder with minimal gear. Thus, the images will be more predictable and pedestrian. A good photographer will show up with the lighting equipment and other accessories necessary to do a more professional, compelling image," says Hildreth.
- **Seek realism.** "I try to be diligent in making sure what I am having my subject do is as realistic as it might be within those particular confines," says Hildreth. "We don't want the subject to do something they would never do, but perhaps something they might do." For example, says Hildreth, he would not shoot a senior scientist who does not do bench work wearing a lab coat and goggles.

Protect Your Photo Rights

If you hire a photographer, specify that you are hiring on a contract basis and will require unrestricted use of the resulting images. Otherwise, the photographer will hold the rights and will charge you for each use. You will pay a photographer either for a half-day or full-day shoot. The photographer will not come cheap. An hourly fee of $100 and up is not unusual.

If your research is done under a contract with a foundation or other group, make sure you understand your photo rights under that contract. For example, a grant from the National Geographic Society involves signing away many photo rights that could compromise your ability to communicate your research. "I have spent considerable time telling faculty members 'No, you cannot sign that piece of paper," says Ohio State public information office (PIO) Earle Holland of such cases. "That piece of paper says you can't use your own images for academic publication. You are signing away your rights to the best images to receive a $15,000 grant, when you have a $500,000 research program.'"

When in doubt about photo rights, ask your PIO, your contract office, and your in-house photography office to review any contracts, because each can offer different expertise.

Prepare for the Shoot

You will prepare for a lay-level photo shoot differently depending on whether it is for a news story or a feature. For any shoot, a photographer will want to scout your laboratory. A magazine feature photographer will likely do a location scouting the day before a shoot, but a newspaper photographer will only show up the day of the shoot. In the scouting session, give the photographer any news releases or feature stories about your work, as well as the URL of your Web site—which no doubt contains an extensive, brilliantly written lay-level explanation of your work.

You or somebody else who knows the work should show the photographer visually interesting elements of your work. These include important pieces of equipment, experimental procedures, and striking computer images. Once you have an idea of which areas of the laboratory will be shot, make sure they are tidy.

In planning shots for media release, keep in mind that ideally the shot should feature one person and a visually interesting piece of equipment or activity. The shot will lose impact if two people are featured, and may be rejected by media if three or more are shown. However, for political and/or credit reasons, you may want to have more team members in a media shot. If you absolutely need to have multiple people, the photographer might still come up with a usable shot by placing one or two of the most important people in the foreground, with others engaged in a research activity in the background. Or, the photographer might take two shots—one for media outlets showing a couple of people, and the other including the full team that can be published in internal media and posted on your Web site. The least desirable option is for the senior researcher(s) to simply bow out of the image, giving junior people the spotlight. This option is undesirable because the senior researcher will likely be quoted in the release, and media will expect the photo to depict that spokesperson.

If your work involves animals, make sure you observe any institutional policies regarding lab animal photography. Plan very carefully with the photographer any shots involving animals, so that they will not be misconstrued. For example, even though monkeys may be very comfortable in restraint chairs, you run a major risk that any such image will be perceived as torture.

Creating the right shot for a target publication is a crucial part of planning a shoot. For example, professional science magazines might prefer simply a shot of the animal or equipment involved in the research, without a human. Conversely, popular magazines will definitely want a dynamic laboratory image that includes a human.

If the photo shoot involves taking both technical and lay-level images, discuss which technical images you will need and how they will be integrated into the shoot with the lay-level images.

You should come away from the scouting session with a firm idea of which shots will be taken and where. Also, the photographer can tell you how shots involving you will be set up to minimize your time requirement. Usually, the photographer can use a stand-in to establish lighting and camera angles before you have to be involved.

The shoot itself will take about an hour for a news story and as much as a day for a magazine feature, so plan accordingly. The quality of the images will be reflected in your time commitment, and those images will likely have a very long life.

Ask Questions First, Shoot Later

During the shoot, since the photos portray your research, you ultimately control the shoot when working with in-house or contract photographers. And although you do not have complete oversight for media photographers, you can still heavily influence the shots that are taken. So feel free to "art direct" a shoot by asking to see digital images in the camera as they are shot. And if the shoot is on film—as high-end photography sometimes still is—the photographer will usually take a Polaroid image, which you can also review.

As you review images, keep in mind a problem many researchers have with lay-level photo shoots: they are concerned more with what their peers will think than about how effectively the images will connect with their audiences. While you should not accept images that are inaccurate, give the photographer some creative room. He or she will often come up with visual approaches that you have not even thought of. Remember that the purpose of a lay-level image is not necessarily to portray the substance of the research, says Hildreth. "An image needs a creative way of visually hooking a prospective audience into reading a story about their research," he says. "Without that visual hook, people will glance at the image and move on. But an interesting image stops people, makes them study it, lures them into reading the caption, and then you've hopefully got them into your story."

Nevertheless, you might feel a photographer is "overshooting" an image—for example, adding elaborate colored lighting effects or odd angles to make a shot more sensational than you would like. Avoid such overly theatrical techniques, especially for a news shot. However, for feature stories, in which images aim to make an editorial statement, more elaborate staging may be appropriate. If you are not comfortable with such effects of composition or lighting, discuss your qualms with the photographer and agree on a shot with which you are comfortable.

Understand the Review Process

A photo session produces a myriad of images from which a final few will be selected, and you should understand ahead of time how that selection will be made. For simple news photos, an in-house photographer might make an initial selection for you to review or even to select the final shot. For more complex shoots, you will want to collaborate with the photographer and the PIO to make the selection. In this review, you contribute your perspective on what works scientifically, and the photographer can point out composition and aesthetic issues that will affect the decision. And your PIO can offer input on how the images communicate the essence of your work.

Of course, you have no control over photo choice for shots by newspaper or magazine photographers. So take special care during the shoot to avoid shots that you would not want published.

Think Web

As you review possible images, keep in mind that they will almost certainly be used on the Web, which affects both composition and color. "Images are played smaller, so often subjects should be shot tighter," says Hildreth. "Otherwise visual impact is lost. Also, contrast and color on the Web is a crapshoot. They will change from monitor to monitor. So, a photographer has to plan contrast so it is in the middle. If a shot is too contrasty, darks are too dark, and highlights are too light on the Web. And if an image is too flat, it looks muddy on the Web."

The best way to ensure that Web images are displayed with maximum visual impact is to post them as thumbnail images that can be clicked on to bring up a larger version in a second window, says Hildreth.

Know the Ethics of Photoshopping

Given the power of image manipulation with such software as Photoshop, you will no doubt be faced with issues of how much you can ethically alter both your technical and lay images. In both image types, observe two hard and fast rules:

- **Do not alter an image to change its main subject or editorial point.**
- **If you substantively alter an image for aesthetic or communication purposes, state clearly in the caption what alteration has been done.**

Media photojournalists are strictly forbidden from substantively altering images for publication. For example, perhaps the most notorious case of such alteration occurred in 1982, when *National Geographic* editors used photo editing to move two Egyptian pyramids closer together so that they would fit on a vertical cover.

However, photojournalists have long legitimately altered images to improve quality. "There are technical changes that deal only with the aspects of photography that make the photo more readable, such as a little dodging and burning, global color correction and contrast control," wrote photojournalist John Long in an essay on ethics for the National Press Photographers Association. "These are all part of the grammar of photography, just as there is a grammar associated with words (sentence structure, capital letters, paragraphs) that make it possible to read a story."

In contrast to media photography, photojournalism guidelines are less stringent for images produced in-house for research communication, Hildreth notes:

> We do everything we can to create the image in camera. But we are not doing photojournalism. This is public relations, and we have a little more latitude than if we were doing pure photojournalism. For example, if say a soda can was overlooked during the shoot, and we don't want to be doing a product endorsement, Photoshopping it out would be fine. But I would probably draw the line at inserting an image into a video monitor that is in a shot and that wasn't working that day. We would rather come back and reshoot. And clearly, we wouldn't alter images to move objects or people.

NASA's routine use of false color and compositing in space images is a prominent example of image manipulation for communication purposes. In such cases, NASA clearly indicates in captions how false color or compositing was used and why it contributes to communicating information in the image.

There are also instances in scientific publication in which altering photos is permissible. For example, there is the case of the cover image produced by

Howard Hughes Medical Institute investigator Charles Zuker for an article in the August 10, 2001, journal *Cell.* The journal was publishing a paper by Zuker and his colleagues on the functional identification of the sweet-taste receptor. A cover image would serve to dramatically highlight the research achievement, so Zuker hired a professional photographer to shoot mice nibbling at a luscious-looking chocolate pastry. Of course, persuading mice to pose perfectly was impossible, so the photographer shot the mice and pastry separately and Photoshopped the animals into the image. Given that the image was clearly meant as cover artwork, rather than scientific data, it was perfectly acceptable to alter the image.

Produce Personable People Pics

The hallmark of an amateurish Web site is the posed group shot of team members, all blandly smiling in a row, as discussed in chapter 7. Such photos not only bore viewers but also present your research team as a rather anonymous collection of faces, rather than individual talents. Also, group images go out of date the instant somebody leaves or arrives. It is far better to create a gallery of individual images of lab members, along with their bios. In such galleries, Hildreth recommends against the standard head-and-shoulders shot. Rather, images should show members of the research team at work, doing what they normally do, he advises.

And while you or a colleague might take perfectly acceptable lab member shots, have a professional take the photos if you have the budget. A professional photographer can light the subject in a much more attractive way. Also, to be blunt, the photographer can arrange a shot to minimize that extra chin, beaky nose, or fright-wig hair. Some tips on such headshots:

- **Your headshot should look like you.** Avoid elaborate lighting effects or poses that do not capture your true features. You want your readers to recognize you. So forget the soft focus, alluring expression, and feather boa.
- **Stay recent.** Renew your photo every few years, especially if you change your look. A ten-year-old photo is both confusing and unprofessional.
- **Wear simple clothes or jewelry.** Solid color clothes and conservative accessories such as ties or jewelry work best. Gaudy accessories distract from your face and compromise your professional image.
- **Use modest makeup.** For women, makeup should be basic. Men can ask the photographer whether powder is necessary to reduce shine, particularly on a bald pate.
- **Take multiple shots.** Varying angles, poses, and facial expressions will give you a selection for different purposes. Some photos should have a plain backdrop so your Web designer can knock it out if necessary.

Make the Most of a Studio Shoot

News shots might also take place in the studio, in which case you should consider bringing props or images for use in the shot. Consult with the photographer about the best materials to bring and how they might be used.

A good "two-fer" approach for such studio portraits is for you to pose in front of a slide or computer image depicting some visually interesting aspect of your work—an animal, a colorful graph, etc. The resulting image will serve both as a visually interesting portrait of you and an illustration of a concept central to your research. In making such images, keep in mind that such a portrait may be used to illustrate more than one news release. So, it might be a generic shot of you with a piece of equipment or image that you use throughout your work. However, notes Hildreth, a laboratory shot is still preferable. A studio research shot is usually a fallback strategy when research has already been done or was done in the field.

There may be instances that a graphic critical for explaining your work lacks visual interest, in the photographer's opinion. If so, be prepared to ditch that precious graph or chart, or rework it to be more accessible and visual. If you absolutely must have a graphic in the shot that does not work for a media image, arrange to have two photos made—one with the graphic and one without. Your too-technical shot can be used for internal publications and Web sites.

Embed Your Photographer

To a PIO, one of the greatest disappointments is a researcher who participated in an exciting, productive field expedition but brings back only amateurish photos as a visual record. Such expedition photos usually include boring shots of researchers lined up in front of a tent or the tops of people's heads bent over some instrument.

Professional-quality images from an expedition will enormously benefit the communication of its scientific findings—not only to lay audiences but also to scientific audiences. So, consider taking a professional photographer along on an expedition or inviting a photographer to a day of field research close to home. You will likely be surprised at how useful such professional field images are. They will enhance your presentations and lectures, news releases, your lab's and institution's publications, and even research proposals. They will also provide a far richer visual record of your research, perhaps capturing a Eureka! moment you had not realized at the time.

Even if you or other expedition members are adept photographers, a professional photographer is a better option, emphasizes Hildreth. "Having an individual dedicated to the visual record of the expedition allows a researcher to not have to worry about or think about that task," he says. For example, Hildreth has descended into Madagascar caves to record the search for remains of ancient lemurs. He has trekked the jungles of Sumatra to photograph orangutans. And he has taken aerial photography of the gigantic Arecibo radio telescope in Puerto Rico.

An expedition photographer should already have such field experience, advises Hildreth. Field photography requires more than the usual photographic skills. It requires the logistical abilities involved in organizing and transporting masses of equipment in less-than-hospitable environments. It also requires personal and political skills involved in adapting to often exotic local customs and negotiating for the best shots. It can also require personal stamina. On his Madagascar expedition, for example, Hildreth had to lower loads of equipment and himself on a single rope down many stories into a cave. He had to figure out how to light shots of paleontologists in the vast blackness of a cavern. And he had to persuade local tribesmen to pose for his shots.

To benefit optimally from a field photographer, spend considerable time briefing the photographer on expected conditions and research activities. Discuss the kinds of images you want and how you plan to use them. Involve the photographer in planning meetings, so that he or she can integrate his logistical needs into yours. Also, listen to the photographer's suggestions of adjustments that will enhance the visual record—for example, planning an experiment for a time of day that photographs better. A good field photographer will not only look for good shots but also make good shots happen. Without compromising the scientific value of the expedition, try to accommodate the photographer. The result will be a better photo that more effectively captures the research.

Even the most extensive planning might not be enough, though. In the field, both you and the photographer should take advantage of surprise opportunities for good shots, says Hildreth. "The researcher and the photographer have to partner in creating the images," he says. "There may be things the photographer sees based on his skills that the researcher is not keyed in on, and there may be things that the researcher understands and sees that the photographer isn't keyed in on."

Funding for an expedition photographer need not come entirely out of your pocket, but can be shared. Grants often contain stipulations that some funds may be used for communications. Also, the funding agency might be willing to authorize supplemental funding, in return for using the images for its own purposes. Photography costs might also be partially underwritten by participating corporations, in-house publications, the development office, the news office, or other such units. What's more, the travel costs of an expedition might be shared if the photographer obtains other assignments in the region. For example, for a trip to Sumatra to shoot Duke research on orangutan communication, Hildreth obtained funding from the alumni magazine, the news office, and other offices. He was also funded by the international study office, since he stopped en route in Paris to shoot images of Duke students studying abroad. In such cases, funding participants were offered either free use of images or a considerable discount on the use fees, in consideration for their support.

Create Online Image Galleries

When planning photography, consider creating online image galleries, both technical and lay-level, to portray your work. A gallery enables you to tell the story of your research in an engaging visual way that text cannot. Also, a still image gallery is easier to create and maintain than a set of videos—although by all means use video if it more effectively portrays your work. Galleries can also provide an excellent visual portal to important information on your Web site. For example, you can include links in an image caption to information on the machine or concept portrayed. Resources for this chapter at *ExplainingResearch.com* feature links to some exemplary galleries.

Doing a gallery need entail no more than creating a Web page of thumbnail images that link to larger images and captions. However, it is almost as easy and much more professional-looking to produce a gallery that is a more self-contained experience, occupying its own window and with a theme and continuity. Images in a professional-quality gallery can even be made to "move," by choreographing scans and zooms to emphasize key elements. Such galleries can also feature different transitions between images and elaborate presentation of text, buttons, and sliders to enable viewers to control the presentation.

Professional software for producing multimedia shows includes 3-D Album, Microsoft Expression Media, QuickTime Pro, and Adobe Flash. LiveSlideShow creates QuickTime movies from slides. If you are not familiar with multimedia tools, Soundslides enables you to create audio slide shows with no need for training. Articulate Presenter is a more elaborate e-learning software that also includes slide shows. The online resources for this chapter at *ExplainingResearch.com* include links to all these tools. Also, your webmaster or Web services office can help in choosing the software and developing the gallery.

In developing your gallery, you obviously need good images with strong composition and interesting subjects. Avoid, for example, a gallery that is nothing more than a series of people posing stiffly with machines. Rather, the images should include interesting and varied angles and lively subjects. If you are starting from scratch, first convene a meeting that includes the photographer and designer to discuss the vision of the gallery and to work out the images to be shot.

Your gallery should have a unifying theme or story. The gallery can consist of a set of related images that combine to help viewers understand a piece of research. Or, it can be a sequence that carries the viewer through a process or idea. To convey complex or abstract concepts, you can even intersperse the photos with diagrams that explain the research. The images should generally be horizontal, given the dimensional ratio of computer screens. They can also be vertical, but avoid mixing the two, cropping images to be one way or another.

An audio narration gives a gallery a more engaging, personal feel. For example, the *New York Times* does an excellent job with such galleries. Its multimedia page also showcases video and audio multimedia features, including many about science. Your gallery should include no more than a dozen or so images, and its narration should not go over five minutes. Thus, each image should have no more than about 30 seconds of narration. Also, for narrated slide shows, keep the captions simple. Simpler captions allow viewers to engage themselves in the images and sound without trying to view images, read the caption, and listen at the same time.

For narration, "amateurs" such as scientists often prove more engaging than professional announcers. Even though a pro would clearly give more polish to the gallery, researchers talking about their own work is more involving and immediate. An exception, of course, is if you have a thick accent or a distinctively unpleasant voice. Ask the radio/TV director in your news or multimedia office to give you a bit of assessment and coaching. With coaching, preparation, and editing, you can produce a good narration even if you have not done any radio work before. Some tips:

- **Rather than preparing a script, develop a bulleted list of points to make for each image. An amateur reading a script invariably sounds stiff and awkward.**
- **Do not worry about being perfect in recording the narration.** You can always repeat and rephrase during the recording session. It is best to rehearse a few times to get the verbal marbles out of your mouth.
- **Take advantage of sound engineers' absolutely magical editing tools to tweak your narration.** Using software such as Adobe Audition or Pro Tools, a sound editor can edit out pauses, stammers, and uhs. An engineer can also adjust bass and treble to make even a squeaky voice sound good.

Finally, an engineer can incorporate music or environmental sounds to enrich the presentation's impact.

- **Eliminate background noise.** While the best venue for recording a narration is a sound studio, perfectly acceptable narrations can be recorded in a quiet room using a digital recorder, a minidisk, or digital audio tape recorder.

Create 360-Degree Shots

If your research involves fieldwork or other visual subjects that lend themselves to immersive images, consider producing 360-degree images for your presentations and Web site. The two most popular systems for making such shots are IPIX and QuickTime. Another, more elaborate system is GigaPan, which requires special equipment for high-resolution panoramic images. See resources for this chapter at *ExplainingResearch.com* for links to information on these systems.

Of course, panoramic images will require a professional photographer with the right equipment and software. However, real estate companies now routinely offer 360-degree virtual tours of houses, so those commercial services are readily available for producing tours of your laboratory, field site, or other facility.

If you are at a university, you might find advice on creating a virtual tour of your laboratory right on your campus. Many universities have created virtual tours, accessible through *CampusTours.com*, that include panoramic views. One excellent example is Harvard's virtual tour, which includes 75 linked QuickTime panoramic images.

13

Produce Informative Research Videos

Given the ease of creating and editing video, you should strongly consider including video on your Web site and in your presentations. Videos depicting your work can offer significant insights not possible with text, audio, still images, or animations. For example, video can better explain laboratory procedures than can text, as shown by the videos on the *Journal of Visualized Experiments* at *JOVE.com*. In fact, such videos can aid acceptance of your findings by making it easier for other researchers to reproduce the procedures involved.

Video can also add an important explanatory dimension to your papers and posters. For example, the Web site *SciVee.tv* enables researchers to post "pubcasts," which are video explanations of their latest paper, as well as "postercasts"—video explanations of their posters. *SciVee.tv* also enables users to upload lecture and conference videos.

Besides their explanatory power, videos can lend a sense of intimacy and personality to your work that engages people and thus aids your research communication. Video interviews with researchers talking about their work humanize them, says Vanderbilt Public Information Officer (PIO) David Salisbury, who produces such video profiles. "We aim to show who these people are," he says. "They are extraordinary people. Their motivations are like that of hunters—the thrill of the hunt—and like detectives trying to solve a fascinating puzzle that takes all of their wits and concentration."

Your videos need not be slick and elaborate to be interesting to professional or lay audiences, or even the media. While professional production values are a plus, the real key is interesting content. For example, University of California–San Diego PIO Kim McDonald has found media interest even in technical videos done by self-taught researchers. "One guy who was a video buff had made QuickTime videos of nerve cells," says McDonald. "His video looked so unique that we accompanied a news release on his work with it. We had TV stations come to my office and shoot the video running on my monitor. And they used it on TV. So I became convinced that much of the video that labs do on a routine basis can be fine for even the broadcast media."

Consider the Entire Video Spectrum

Your videos can range in quality from simple Web videos for your colleagues to high-definition broadcast-quality "video news releases" (VNRs) for media. And your equipment may be cell phone cameras, webcams, inexpensive pocket cameras, high-end consumer cameras, or professional cameras.

In planning a video communication strategy, consider how all of these might be useful. A cell phone or pocket camera might be fine for a quick interview with a poster presenter whose work interests you—which you can show back home in a lab meeting. On the other hand, only a professional-quality camera will do justice to your work in a VNR. The rest of this chapter emphasizes high-end consumer and professional-level video, but remember that quick, inexpensive video can play a useful role in both professional and lay-level research communications.

Quality Always Counts

The reality is that regardless of your budget or expertise, your audience will judge your video alongside professional-quality videos. So, your video should be as good as you can make it, whether a Web video or broadcast-quality VNR. Fortunately, you can make quite good videos yourself on a limited budget. Also, your institution may have a video production office that offers production services at a reasonable price. Or, you might enlist a student from a video production program. While a student's video might not be as good as that produced by a professional, the cost will be lower and the video might serve your needs perfectly well. Even using a professional videographer is not all that expensive. For example, the *Journal of Visualized Experiments* charges about $1,000 for professional production of a video for its site, which is comparable to journal page charges and to the costs of producing scientific graphics for a paper.

To get a good overview of the production process, see the tutorials and books listed in the online resources for this chapter at *ExplainingResearch.com*. Mark Pope's guide, in particular, usefully discusses the various roles: the producer, writer, videographer, graphics specialist, director, talent, voiceover, and editor. Of course, in a small video production, one person plays many of these roles. Pope also describes the organization of the process into

- Preproduction: information gathering, scriptwriting, talent selection, and location selection
- Production: shooting video, shooting voiceover, digitizing, graphics creation, motion effects, and music and sound effects
- Postproduction: rendering, editing, and output

If you decide to produce your own video, besides asking for advice from your video production office, consider taking a videography course at a local school or in your journalism school if you are at a university. Even if you plan to use a professional video production service, taking such a course is worthwhile to get a basic idea of how to work with professionals and what to expect.

Observe the Basics

For any type of video, observe these basic principles:

- **Target your audience.** While there is certain information you want to convey, also ask yourself what your audiences are interested in knowing and how to best tell your story to attract them. Revisit the lessons on understanding your audiences in chapter 1.
- **Plan visually and thoroughly.** Observe the dictum in the television industry, "Say cow, see cow." That is, any concept you are explaining in your video must have an accompanying visual. So once you have decided on the points you want to make, brainstorm how to portray each point visually. You may already have quality graphics or animations you can use or adapt, a piece of laboratory equipment that can be made visually interesting, or an especially visual aspect of the research process that you can capture. Such planning is especially important because video shoots are more complex than photo sessions. They must contend with motion and audio, in addition to composition and lighting. By planning thoroughly, you will not bedevil the crew with unexpected complications.

- **Plan your format.** If you plan what format you will need to output, for example high definition, you can shoot at that level throughout the production. This format planning will depend on how and where you expect the video to be shown.
- **Take freebies.** Give your video a polished look and increase interest by incorporating free video segments available from funding agencies and other sources. VideoUniversity offers a comprehensive guide to public domain footage, and NASA offers a gallery of video segments about its projects. Your in-house video production office also will likely know of other good sources.
- **Script tightly.** You may want to script the video, as described below, or if it is an interview video, work from a list of questions and answer points. Either way, remember that video requires tighter writing than print. Whether for a technical or lay video, edit your verbiage down to an efficient minimum. Practice reciting your interview answers or script text until the words roll trippingly off your tongue. Your PIO or another independent audience can give you feedback. For simple videos, your planning can be informal, and for more elaborate videos, you can use commercial software such as Celtx to organize all aspects of preproduction, including writing the script and organizing the shots.
- **Perform for video.** Tips on giving a TV interview are covered in chapter 24. Read it before you shoot your video, so you will know how to use gestures, inflection, and energy to convey your information effectively.
- **Synergize your shoots.** As long as you are shooting one type of video—whether a technical, news, or field video—plan the shoot for as many purposes as feasible.

Check Out the Good, the Bad, the Ugly

Watch some science videos to get an idea of what is good and bad. To see some of the best in science videos, explore the segments of Nova ScienceNow. In particular, view their Dispatches—short reports from producers and correspondents. These segments use all the above principles to create informative, engrossing videos. To see a broader range of good—and very bad—science videos, go to YouTube and search on such terms as research, physics, biology, and so on.

The online resource section of this book lists many such sources of well-done news, interview, demonstration, and lecture videos about science, engineering, and medicine. These sources include the Australian Broadcasting Channel, AthenaWeb (European science videos), the Discovery Channel, the Honeywell

Nobel Interactive Studio, IEEE Spectrum videos, National Geographic videos, the Research Channel, SciTalks, and Wired Science videos.

For examples of excellent university-produced science video demonstration/lectures, see the University of California–San Diego programs Science Matters (life sciences), Atoms to X-rays (physical sciences), and Grey Matters (neuroscience). Also, see the university's Molecules for the Media press workshops in the physical sciences. These represent particularly effective examples of educational science videos produced by a university news office. They were produced by UCSD PIO Kim McDonald and colleagues in his office and at UCSD-TV with corporate and foundation support. By enlisting outside funding, they effectively overcame the endemic problem universities have with supporting video projects.

Create Informative Technical Videos

In creating technical videos, keep them short, no more than five minutes. Viewers may be perfectly comfortable watching an hour-long documentary on TV while lounging in their favorite chairs. But sitting hunched over a computer screen watching a long technical video would be excruciating.

If your research requires more time to explain, consider breaking the video into topical chunks of a few minutes each. This parsing will make the videos both more digestible and more accessible—in that viewers can choose which topics to address. Also, such chunking enables you to adapt segments for use in your presentations.

You might believe that because you are making a technical video, you can get by with lower quality videography, editing, and such. However, high-quality technical videos prove more effective at engaging the professional audience you wish to reach. And even the most tolerant professional audience will unconsciously compare your video to the commercial video they are used to watching.

What's more, even the most technical of videos will invariably reach lay audiences, including students and your institution's administrators. Do you really want such audiences seeing an ill-lit, poorly edited, poorly narrated video of your work? So, if you do not have the resources for a professional videographer, make sure you acquire the skills to make quality videos yourself.

Make Dynamic News Videos

News videos have a more constrained format than do technical videos. A VNR meant for television stations can run no more than 90 seconds and must have a succinct introduction, sound bites of no more than about nine seconds, and quick visual cuts. The VNR usually includes a narration on an audio track separate from the ambient sound, so TV stations can insert their own narration. Such videos are usually produced by your news office for distribution to commercial stations and posting on news Web sites.

"B roll" is a more broadly useful video package for commercial stations than a scripted VNR. Such B roll video comprises a collection of scenes that the station can assemble into its own video. The B roll is accompanied by a shot list describing the scenes and a copy of the news release behind the video.

While some news stations may use the VNR or edit the B roll into their own story, others will likely do "readers," in which the news anchor reads a ten-second summary of the story while a brief video segment shows.

Given their brevity, VNRs require more distillation of concepts than do print releases, although you should still be as accurate as you are in developing news releases. You can allay any frustration over the necessary brevity of VNRs by thinking of them as "video headlines" that will attract interested people to the richer explanations of your work in your news releases and on your Web site.

Avoid Video News Ethical Pitfalls

An insidious problem is that some VNRs constitute little more than promotional material masquerading as news. For example, hospitals sometimes make deals with local TV stations to run their promotional videos as "health news" when the segments are actually only advertisements for the hospital. In an article in the *Columbia Journalism Review*, author Trudy Lieberman calls such arrangements

> the product of a marriage of the hospitals' desperate need to compete for lucrative lines of business in our current health system and of TV's hunger for cheap and easy stories. In some cases the hospitals pay for airtime, a sponsorship, and in others, they don't but still provide expertise and story ideas. Either way, the result is that too often the hospitals control the story. Viewers who think they are getting news are really getting a form of advertising. And critical stories—hospital infection rates, for example, or medical mistakes or poor care—tend not to be covered in such a cozy atmosphere. The public, which could use real health reporting these days, gets something far less than quality, arms-length journalism.

In the interest of ethics and long-term credibility, avoid such relationships, as tempting as they may be.

Another ethical problem arises when corporate research sponsors offer "educational grants" to support VNRs. "This is not a pot of money they're trying to give for education," says Don Gibbons—formerly communications officer at Harvard, and now with the California Institute for Regenerative Medicine. "Its aim is to support producing VNRs through an agency for PR purposes. Administrators are all too often taken in by the ploy," he says. Gibbons recalls vetoing such a grant when a department chairman had already accepted it. "This person thought the grant was a great opportunity to inform the public, so why not take it?"

Should You Produce a Lecture Video?

When considering posting your talk on the Web, your first instinct may be to record a video of it. However, your talk really warrants a video only if it is particularly visual, for example, involving demonstrations. In such a case, you could create a straight streaming video or use high-end systems such as Echo360 or Sonic Foundry Mediasite to create online multimedia packages, as discussed in chapter 14.

However, if the video will only show you as a talking head, it is probably not worth the expense; after all, video lectures often require multiple cameras and considerable editing. A less expensive and perfectly accessible alternative is to produce a narrated slide show, as the New York Academy of Sciences does with its eBriefings. The academy records the talks and uses Articulate Presenter to produce narrated Flash-based lectures from PowerPoint slides. Or, as mentioned in chapter 3, you can also produce a slidecast using the SlideShare or SlideServe services.

Capture Field Research on Video

Video of your field studies can offer the most compelling view of your research. Consider embedding a videographer in your fieldwork, especially if you need high-quality documentation. However, hiring a videographer for the field is usually difficult, so with coaching, you and your colleagues can produce your own video. For example, Ohio State PIO Earle Holland helps his researchers develop both their video and photography skills. "I will volunteer both our still photographer and videographer to review images of their past expeditions and tutor them on such things as shot composition," he says. "We can tell them things they may never have thought of. For example, we suggest they shoot a scene once for documentation with the time code on, and then turn off the time code and reshoot so they have a clean shot for media purposes." Holland has also created a system for adapting field video to media use.

> For example, when glaciologist Lonnie Thompson goes to the Antarctic, Kilimanjaro, the Himalayas, or other places, he takes high-end digital still and video cameras. He produces many hours of video, and when he gets back, I have our videographer go through it and grab twenty or thirty minutes. Out of this, we make a media tape, so when a story comes out, we not only offer stills, but also video. And that video has been used worldwide, by American networks, BBC, Japanese TV, Italian TV—the list is endless.

Write Your Script

Writing a video script to be recited is quite different from writing prose. For example, if you read transcripts of ScienceNow segments, available online, you will be surprised at how terse, even simple-minded, they seem. Such terseness is necessary because a video engages two senses at once—visual and auditory. So, overly-complex verbiage is useless noise, distracting the viewer from the visuals and thwarting communication. Here are guidelines for writing a compelling, concise video script:

- **Write it tight.** Write short declarative sentences, no more than 15 words, and using simple words. Shorten titles. Leave out technical terms. Simplify quantities by saying "almost 600," rather than "589."
- **Brighten it.** Grab your viewers by using humor, phrasing, personality, and energy. Take a lesson from commercials, the most efficient visual storytelling medium ever invented. They are stunningly effective at using such elements to engage an audience. However, avoid clichés and corny jokes.
- **Recite it.** Once you have a tight draft, recite the script out loud, paying attention to words and phrases that may be garbled when spoken. Keep reciting and tightening until the script says what you want, yet flows smoothly. Recite the script to a test audience—either volunteers or those bribed with cookies or other treats. Ask them for frank feedback on whether the script tells the story accurately and engagingly.
- **Retighten it.** Write and recite several drafts, putting aside each draft for a time and coming back with a fresh eye (and ear) and a needle-sharp editorial pencil. Ruthlessly edit down complex verbiage and long sentences.

- **Time it.** Once the script is final, time it to make sure it is not too long—a technical video should run no longer than about five minutes, and a news video 90 seconds. Make sure the length of time that you are discussing something like an instrument matches the length of the scene depicting it. Also, your script must allow visuals to be onscreen long enough for viewers to make sense of them, usually 2 to 10 seconds.

Shoot Your Video

If you have the budget, use a professional videographer, or at least a talented student, to shoot your video. The expense is well worth it, especially for video that will be widely seen. However, if you do shoot the video yourself, besides taking the tutorials listed in the online resource section, observe these tips:

- **Use a quality camcorder.** Preferably use one that records on miniDV tape. Tape is less expensive and has greater capacity than do the DVDs in DVD camcorders. And tape can be archived more easily than the hard drives or memory chips in those camcorders. Your camera should have a quality optical zoom, image stabilization, and an external microphone jack. Also, choose a high-definition camera. Even though most people will see your video in lower definition online, a high-definition version is useful for presentations.
- **Storyboard your shoot.** Plan your shots by sketching the shoot as you would a series of comic book panels, with each shot matched to an element of the script.
- **Take control of the environment.** Before the shoot, do not hesitate to rearrange equipment, people, or other components to achieve the look you want. And, as indicated below, take control of lighting and sound.
- **Think lighting.** Do not settle for existing lighting for your video. Overhead lights can cast shadows on your subject. And overhead fluorescent lights can give your subject a green tint. Either arrange your subject to take advantage of good natural light, or use "three-point lighting" if possible. Such lighting consists of a direct light to one side of your subject, a reflector on the opposite side to fill in shadows, and a small backlight to highlight the subject from the back. You can light cheaply using a tungsten work light from a hardware store, and for fill light use a silvered automobile dashboard reflector. Do not mix different types of light sources. Finally, set a "white balance"—the camera adjustment that defines the color of pure white—for each shot to ensure consistency of color.

- **Think sound.** For interviews, minimize environmental noise by turning off equipment, closing doors, and so on. Avoid using your camcorder's onboard microphone. It picks up camera sounds and your inadvertent mutterings and throat clearings, besides the sound of your subject. Instead, use an external lavalier microphone, which you can clip to the lapel of an interview subject or position near a sound source. A wireless microphone is even better, although it is more expensive. Monitor the audio with headphones to be certain you are getting quality sound. Also, if there are other sounds integral to your story, such as an animal or instrument sound, make a quality recording of that sound.
- **Shoot to edit.** Shoot a wide variety of wide-angle and close-up shots to choose from in editing. It is difficult to go back and reshoot, so when in doubt, shoot it.
- **Shoot establishing shots.** These are broad shots that establish the overall location. For example, if you are shooting a video about work in a laboratory building, start by shooting the outside.
- **Shoot cutaways.** If you are interviewing someone, as that person is speaking you want to be able to cut to relevant shots. Just showing the interviewee can be boring and less informative. Shoot anything that the interviewee is discussing. And if something catches your eye—a test tube sloshing or a digital counter blinking—shoot it. You might want to use the shot as a cutaway.
- **Shoot short scenes.** As mentioned above, typical scenes last from 2 to 10 seconds. Longer shots bore the viewer.
- **But also shoot fat.** That is, start the camera 20 seconds or so before the action you want to capture, and stop it a few second after. In editing, you will be thankful for these heads and tails.
- **And frame fat.** Leave some headroom at the top of the frame, and leave space anywhere you plan to insert titles. Do not necessarily center your human subject in the frame for an interview, but perhaps a bit to the left or right.
- **Shoot steady.** Stabilize stationary shots by installing the camera on a tripod, resting it on your shoulder, or bracing it against something solid.
- **But also move it!** Carefully done motion can make your video more dynamic. For steady tracking shots, hold the camera at your waist or use a "steadycam," a device that stabilizes the camera as you move. The online resources for this chapter at *ExplainingResearch.com* list plans for homemade steadycams, stabilizing devices, cranes, and other useful gadgets. You can make smooth dolly shots by having someone push you

along in an office chair or wheelchair as you shoot. Or, you can shoot from a moving vehicle with air let slightly out of the tires.

- **Avoid pans and zooms.** They are unnatural and distract from your video. Your eye does not pan or zoom, so why should your video?
- **Minimize movement for Web video.** For low-resolution YouTube or Web site videos, stick to stationary shots or slow movement if you must move at all. If your video will be shown on a large screen, consider shooting alternative scenes with movement.
- **Think background.** The background of your subject should be either darker than the subject or out of focus. Do not shoot your subject against a lighted window or a visually busy background.
- **Shut up.** When an interview subject is talking, avoid even a faint "Uh-huh." It will distract viewers and make you sound very dumb.
- **Log your shots.** List all your shots, with time codes, to assist your editing.

Even if you do plan to use a professional videographer, pay attention to these tips and learn basic videography through tutorials such as those listed in the online resources for this chapter, so you can productively contribute to the shoot. Understand the purpose of each shot and what it will look like. One aim of your participation, of course, is to make the video as scientifically accurate as possible. But you might also come up with ideas for scenes that had not occurred to you before the shoot.

Edit Your Video

You will certainly leave the editing of a news video to a professional, but your technical videos will also benefit from professional editing. A professional will have a much more sophisticated grasp of scene selection, transitions, audio, and so forth. However, if you do want to edit your own video, there is excellent software that can be mastered with some effort. And there are good online tutorials available. For example, the editing package Adobe Premier Pro CS4 offers such a tutorial. Other good editing software includes Apple iMovie, Apple Final Cut Studio, Pinnacle Studio Ultimate 12, CyberLink PowerDirector 6, Ulead VideoStudio 11 Plus, and Windows Movie Maker. *PC Magazine* offers a good guide to choosing video editing software.

Even though your video will find the most use online at a lower quality, save your videos as a high-quality MPEG-4 file. Also, consider "watermarking" your scenes with your Web site URL. The URL can appear unobtrusively and with slight transparency at the bottom of the screen and will drive traffic to your site.

Generally, editing of both technical and news videos should be kept simple, with no fancy wipes or other transitions between scenes—only simple cuts. Also, screen text should be kept simple, limited to titles that stress important points. Do not show the same text on the screen as that being said—it tends to insult the viewer.

Create a Web Video

While a professional editor can prepare your video for the Web, you can do it yourself with some training. Media College offers a good tutorial on preparing "streaming" video—that is, video which transmits to the receiving computer as it is played, versus downloaded video, in which the entire video is downloaded to a file and then played. The online resource section for this chapter lists other sources of information on the streaming formats—Windows Media, RealMedia, QuickTime, MPEG-4, and Macromedia Flash. Flash video is by far the most popular streaming format.

Even if you plan to use a professional, having a basic knowledge of Web video will help you make knowledgeable choices. There may be technical issues that determine that choice, or your colleagues may have standardized on one format.

Posting your streaming video will be simpler if you can embed it in your Web page, called "HTTP streaming." A Media College tutorial offers instructions for adding a video clip to your Web page. The other alternative for video posting is a "streaming server," which is usually only necessary if a large number of users will want to view the video simultaneously. Your institution likely operates such a streaming server. So, make the decision based on your projected audience size.

Syndicate Your Video

Fortunately, your video can enjoy a wide audience by being posted on the many Web sites for news and research videos. Your news office can distribute VNRs to television stations and networks, which may use them not only in their broadcasts but also on their Web sites. And even if your VNR is only broadcast on the local TV station, its affiliated network may post the video on its Web site, giving you national exposure. Your news office can also post the VNR on YouTube, Google Video, and other general Web video sites, as well as on the *Scientific American* Web site. TubeMogul offers an easy way to upload your videos to the major Web video sites, as well as to track usage of your video.

In uploading your video, add "tags" that describe the video so it will be found in searches. These tags are key words attached to the video, just as a Web image is tagged. Your tags should include every relevant word that describes your video. Also, some video hosting sites allow you to post relevant URLs in the description of your video. And, of course, when the VNR is posted, you can link to it from your laboratory site—as can your department, school, center, and other research units.

You can post your technical videos yourself, to such outlets as Bioscreencast, DNAtube, doFlick, Journal of Visualized Experiments, LabAction, ScienceHack, and SciVee.tv. TeacherTube posts instructional videos, and the Liberated Syndication service hosts general videos. You can also post short technical videos to general sites such as Google Video, YouTube, and Current.

For longer video lectures, your institution might be willing to feature them on its Web site. For example, the Howard Hughes Medical Institute posts its Holiday Lectures on its site. You can also post long videos of lectures, symposia, and interviews on Apple iTunes U, SciTalks, and Free Science Videos and Lectures. The Research Channel can broadcast your lecture or seminar on cable and satellite services and archive the video on its Web site, if your institution is a member of the Research Channel consortium. Also, the Research Channel video collection is available on Google Video. See the online resource section for this chapter for links to all these sites.

14

Organize Dynamic Multimedia Presentations

You can synergistically increase the impact of your text, audio, and visuals by integrating them into multimedia packages that present your work in a coherent framework. Multimedia presentations can take many forms—for example, virtual tours of your laboratory or center, explanations of a specific research discovery or research program, or multimedia explorations of a topic. They can be relatively simple, such as those produced by The Why Files, PBS's *ScienceNow*, Vanderbilt Exploration, and Harvard's LabWork. Or, they can be quite comprehensive, such as the multimedia presentations of the National Geographic Society, Harvard's BioVisions, *BioInteractive.org*, and the campus tours collected on *CampusTours.com*. The "Tagging of Pacific Predators" Web site illustrates how engaging and interactive a multimedia Web site can be. Created by multimedia pioneer Jane Ellen Stevens and her colleagues, the site enables users to follow in real-time the peregrinations of tagged elephant seals, sharks, and turtles. The site features a photo of the day, researchers' blog, ocean news, ask-a-researcher, and a feature story. It also includes pages for individual animals on the social networking sites Facebook and MySpace.

While creating a multimedia presentation might seem a daunting task, keeping the following steps in mind will enable you to manage the project successfully:

- **Review other multimedia projects, especially those in your field, to understand how they are organized.**
- **Familiarize yourself with video, audio, and other techniques using the resources cited previously in this book.** Also, Adobe offers a good tutorial on creating multimedia projects for students using its software.
- **If your project is too complex to do yourself, identify a multimedia developer; review candidates' other work to see whose approach matches what you want.**
- **If your project is simple, explore software such as 3D-Album that enables you to produce multimedia shows yourself.**
- **Form a development team consisting of the developer, a communicator/writer, and other relevant researchers.**
- **Working with the team, develop specifications for the presentation.** This step includes answering these questions:
  - What is the specific topic?
  - What messages/information do you want to convey?
  - What are the audiences?
  - What do these audiences want?
  - What is the need for this presentation?
  - What outcomes do you want? For example, what is the desired effect on the audience, what do you want them to learn, and what action do you want them to take?
  - Is a multimedia presentation the most effective way to accomplish these outcomes, and why? If so, what media are more appropriate?
  - Will it be only online or distributed on a DVD?
  - What is your budget to meet these specifications, and can your project be scaled to that budget?
- **To review these specifications, form a review committee of representative audience members.**
- **Develop a marketing plan, in consultation with your public information officer, committees, administrators, and others.** Among the marketing steps: news releases, articles in publications, e-mail notification, search engine placement, and requests to Web sites to link to the presentation.
- **Plan content—text images, illustrations, animations, video, and/or audio, including music and narration.** Among the questions:
  - What do you already have, and what do you need to create?
  - Are there already materials available, free or commercial?
  - Have you obtained formal permissions for use and given credit where appropriate?
- **Organize the content by storyboarding the presentation.**

- **Diagram navigation based on the storyboard.**
- **Design the presentation.** It should have a consistent look and feel, including fonts, alignment of components on pages, and use of repeating elements. The design should use contrast to call attention and add interest. This contrast could include lines, colors, spatial relationships, and typefaces. Include a feedback capability and an ability to track usage.
- **Test a design mockup with the review committee.** Does the design resonate with them? Is it attractive without being glitzy?
- **Develop the full multimedia presentation and test it with the development and review committees.** Track how they use it, including what most attracts them, where they are stymied, and where they exit.
- **Tweak the presentation and retest.**
- **Debut the package, and periodically review the feedback and usage for improving and updating the presentation.**

15

Create E-Newsletters, Wikis, Blogs, Podcasts, Social Networks, and Webinars

Your Web site is not the only Internet tool to reach your audiences. E-mail newsletters, wikis, blogs, podcasts, social networks such as Facebook, and webinars can also be important elements of your communication strategy. Such tools are, in fact, becoming de rigueur in explaining your work. Researchers, particularly young ones, expect to see an extensive online presence for both you and your work. If you do not produce such content, you risk becoming perceived as a stagnant intellectual backwater in your field, rather than a vigorous current in its mainstream. So, with the information in this chapter, you can wade right in.

Also, you will still have an inadvertent presence on the Web, even if you produce no online communications other than your Web site. Any online mention of you—news releases, listings in seminar programs, a query to a dog-care Web site about how to treat Fido's worms—will constitute the information that shows up when people search for your name. This information defines your "brand" whether you like it or not. So, managing your brand wisely means ensuring this online information is richly seeded with content that you control.

Reach Out with E-Newsletters

Even the most engaging, dynamic Web site is still a passive communication tool. People must decide to visit it. However, with e-newsletters people need only decide to subscribe, and they will automatically receive information from you, including the latest content from your Web site. What's more, the newsletter's proactive nature engenders positive feelings and a sense of connectedness in your audiences.

An e-newsletter is also the most cost-effective way—in both time and money—to reach out to your audiences. It is easy to produce and distribute, can be made self-subscribing, and requires no printing or snail-mailing. And if you offer useful content, an e-newsletter can prove highly popular. For example, Howard Hughes Medical Institute has more than 10,000 subscribers to the e-mail distribution of its news releases. "We started it because we thought it would be the easiest way to let people know that there is new news to be read on the HHMI Web site," says associate director of communications Jim Keeley. "Now, besides HHMI investigators and people from the HHMI community, our subscribers include journalists, teachers, students, postdocs, members of the public, and researchers from all over the world."

Organizing Your E-Newsletter

First, of course, you should decide whether you are willing to devote the time to producing an e-mail newsletter and whether there is a sufficient audience. Typically, e-newsletters are published monthly, unless they consist of news releases, as with HHMI. A monthly newsletter is frequent enough to maintain a sense of continuity among subscribers, but not so frequent as to be onerous to produce. A bimonthly newsletter, of course, is better than nothing, and if that frequency is all that seems appropriate, by all means publish bimonthly. Each newsletter issue will take perhaps a person-day to produce and manage, which may be divided among several people if others in your group agree to contribute content.

To explore the feasibility of an e-newsletter, form a planning committee of your research group members. The committee should address whether the audience is of sufficient size and importance to justify an e-newsletter, whether there is sufficient content, and whether there is time to do a proper job of production and marketing. The Web site *E-Zine.com* offers a good tutorial on developing e-mail newsletters.

Your audiences might include colleagues, prospective students, administrators, donors, corporate collaborators, and/or the general public. If your audience range is too broad—for example, if the e-newsletter is for a center that serves many constituencies—consider creating different versions of newsletters tailored to different constituencies. Or, the newsletter could be customizable, as are the *New York Times* e-mail alerts and the *Individual.com* newsletters. See the online resources for this chapter at *ExplainingResearch.com* for links to these newsletters.

Your newsletter content will certainly include news about your lab, department, center, school, and so on. However, you might include broader news about your field to make your e-newsletter more useful and attractive and also to give it more impact and authority. Gathering such news is easier than you might think. For example, you can subscribe to EurekAlert! RSS feeds or Google Alerts, specifying the topics of your newsletter as search terms. Such sources will yield a steady stream of news items for your e-newsletter. Also, the social bookmarking sites discussed later in this chapter are another good source of links to articles about your field. These sites include Digg, deli.cio.us, Mixx, Reddit, StumbleUpon, and Yahoo! Buzz.

Also, consider including a personal message or essay in each issue, for example, an update on an important research project or other news from your laboratory or center. Such personalization helps build a rapport with your subscribers.

Editing Your E-Newsletter

If you decide that the audience, content, and time commitment are there for an e-newsletter, here are guidelines for developing your e-newsletter. First, some editorial guidelines:

- **You or another senior person should select the content of each issue, rather than delegating such decisions to an administrative assistant or junior-level researcher.** However, you can use their help to gather candidate material.
- **Keep the text brief and tightly edited, which means it should be written by an experienced writer and not by a student.**
- **As the default, use plain ASCII text so the newsletter loads quickly and is compatible with all e-mail systems.** Many e-newsletters, such as the *New York Times*, do offer HTML and/or Adobe PDF file options that include graphics. For a text e-newsletter, include a hard line break every 60 characters, so the text will not wrap jaggedly with long and short lines when opened by a subscriber.
- **Write a tailored subject line that includes topic and date for each issue, so recipients will recognize it as a new issue.**
- **Specify in the newsletter how often it will be issued and when.**
- **Put the URL of your Web site at the top, so subscribers can easily access it.**
- **List headlines of all features at the top, so recipients can decide whether they want to scroll farther.**
- **Keep the newsletter content simple; do not use large images, all caps, or attachments.**

- **To make the text easily scannable, separate stories and other elements with white space or with lines of asterisks, dashes, or other characters.**
- **Keep URLs short, if necessary using URL-shortening services such as SnipURL or TinyURL.** If URLs run more than 80 characters, some e-mail systems will insert a line break in them, rendering them nonfunctional.
- **Organize each item in the body to have a headline followed by text.** If an item refers to a news story or other Web content, the text should include a brief summary followed by the link to the Web site.
- **Meticulously proofread all text and test all links.** Send preview copies to yourself and others to double check.
- **Include at the bottom of the newsletter a contact name, phone number, e-mail address, physical address, and a note inviting comments.**
- **Include information in each issue on how to subscribe and unsubscribe.** Your software should include an auto-unsubscribe feature.
- **Include an invitation to recipients to pass the newsletter along to others and a link to the subscription page, so pass-alongs can subscribe easily.**

Setting Marketing and Circulation Policy

Some tips on marketing your e-newsletter and managing circulation:

- **Consider sending the first issue of the newsletter to a wide list of people who might be interested.** This mailing will not likely be considered spamming, because you are offering them a useful, free source of information. Also, the first issue should state that unless they subscribe they will not receive further issues.
- **Advertise the newsletter(s) prominently on all appropriate Web pages and in print materials.**
- **Post a subscription form and sample newsletter on your Web site, so people can see what they will be receiving.**
- **Post a privacy policy that reassures subscribers that their e-mail addresses will not be used for any other purpose.**
- **Make it easy for people to subscribe and unsubscribe.**
- **Use a "double opt-in" subscribing system, in which recipients must respond to a confirming e-mail message before they are added to the list.** Once they subscribe, send them a welcome message, inviting comment on the issues they receive.
- **When someone unsubscribes, drop them a note asking if they would be willing to share the reasons why.** You might get some good ideas for improving the newsletter.

- **When you get testimonials, post good ones on your subscription page, with permission.**
- **Archive past issues in a searchable form on your Web site.**

Deciding on E-Newsletter Software

In picking software to manage and distribute your e-newsletter, first check whether your institution offers software such as Majordomo or ListProc for institutionwide e-newsletter services. If you want a more sophisticated system that can also manage discussion groups and post announcements and alerts, consider commercial software and mailing services such as LISTSERV and SparkList. Such software also should enable you to schedule mailings, import and export lists, manage bounced e-mails, and segment lists. This last capability enables you to tailor e-newsletters to specific categories of subscribers. See the online resource section for links to these list management programs.

Use Wikis to Share Information

For those few people who have not yet used Wikipedia, it exemplifies the power of wikis as a tool for creating and sharing information. A wiki is basically a type of Web site designed to enable many people to post and refine content, either publicly or privately. This content can include documents, PowerPoint slides, calendars, images, audio, and video. A wiki can serve as a combination encyclopedia, bulletin board, and news feed. And because wikis include an RSS feed, any new information can automatically be broadcast to the group.

A wiki does not substitute for a well-designed Web page for explaining your research. Nor are wikis as appropriate as blogs for online discussions. However, wikis are useful for sharing information among members of classes, laboratories, centers, and other groups. Wikipedia has even formed Wikibooks, a community for creating wiki textbooks that are constantly refined and updated by users. Wikis are far more coherent and organized for such collaboration than the all-too-common blizzard of e-mails that clog your inbox.

One popular wiki service for business and academe is PBwiki. The online resource section lists other wiki services, wiki "farms" that host wikis, and information sources on wikis. And, of course, you can find a good basic introduction to wikis on Wikipedia.

As a researcher, one good way to get your editorial feet wiki-wet is to check out the Wikipedia entry in your own field and consider whether you have something to contribute to that entry. Is the entry up to date? Is it complete? Is it accurate? If the answer to any of these questions is no, jump in and start editing. Also, you might consider authoring or coauthoring an article on your field for Google's wiki, called Knol, whose articles are prepared by vetted authorities.

An effective wiki should have

- **A clearly defined scope, to avoid having the wiki become an informational grab bag.** A wiki may be organized around an issue, topic, or event, for example.
- **A well-defined community, so that users have a sense of belonging and commitment.**
- **A central facilitator who establishes and manages its structure.**
- **A core of motivated editors who can reliably produce quality content.**
- **A clear commitment that the wiki will be permanent, so that participants will feel that contributing is worth their while.**
- **Introductory tutorials to enable users to quickly learn to post material and other essential tasks.**

Blog Your Research and Expertise

Blogging can be a highly useful way to communicate your research and/or foster online discussion. Blogs are simple Web sites that enable you to post entries in chronological order, with the most recent displayed first. Blog content can be text, images, video, and links to other blogs. Blogs are interactive because readers can post comments, enabling an online dialog. Also, blog software enables key word tagging and categorization of entries, which enables users to search for previous entries by topic. A blog can include news, background, and commentary on a topic, tutorials, reviews of books and papers, and personal experiences. A blog can also be a way to raise money, by including a "donate" button that enables readers to give money through PayPal or another fund transfer mechanism.

ScienceBlogs.com offers good examples of popular science blogs. And *Nature* Network offers good examples of professional-oriented scientific blogs. Also, many university courses use blogs to communicate course discussions and other information. To sample a broader range of blogs, browse the blogging site Technorati and search for blogs by topic on Google Blog Search and *IceRocket.com*. Also, you can have blog posts automatically compiled for you by using a "newsreader" such as Bloglines.

The media have also turned to blogging to give a more informal, accessible voice to their reportage. For example, the *New York Times*, *Popular Science*, *Scientific American*, and *Science News* all have science blogs.

Why Not to Blog

As popular as blogs are, they do have their downside. Blogging takes time. A blog should be updated at least once a week to remain current, and creating a cogent blog post can take hours. Administrators might question the time spent on such an activity—particularly semiancient senior administrators not raised with the Internet. Blogging also puts you in the public eye and could make you the target of critics, since a blog entry is available instantly worldwide. And depending on your station in the research community, you might face sensitivities about the impact of blog posts on key audiences, for example, employers, or donors.

The informal style of a blog might also lead you to venture personal opinions that, while perfectly valid at that moment, can linger on the Internet embarrassingly like a bad odor. Although you can erase a blog post, it will remain in the blogosphere because other bloggers may have referred to it, or it may have been linked to from other sources. Your only real recourse is to append a correction to the original post.

The quick-draw dissemination of research findings on a blog might not be in a researcher's best professional interest, warns neurobiologist/communicator Chris Brodie. "As science communicators we need to be sensitive to not losing what is good about the old system of science, which is that sometimes slow is good. Peer review is a pain but it is not a bad system, and it often works to vet findings. If you have scientists who are blogging, and they are posting their data every day, it presents the problem of science-by-press-conference times ten."

Blogging may make you a reluctant public expert. Especially if you are blogging about health or medical issues, you may receive frequent requests for advice or help from readers, such as patients suffering from a disease. The possibility of such requests for advice means that your blog must have a legal disclaimer that you are not providing medical advice. One example is the extensive disclaimer on the blog of physician Kevin Pho, *KevinMD.com.*

Another problem is that your blog posts may be picked up by the media and interpreted in ways you did not mean. Journalists who write blogs also monitor blogs in their field, and they consider any post, no matter how offhanded, to be fair game for inclusion in their blog.

Blogs also could compromise your career advancement. Tufts University political scientist Daniel W. Drezner described the professional negatives of blogs in an article in the *Chronicle of Higher Education*. Drezner was denied tenure at the University of Chicago, and he cites news media conjecture that his blog was partly to blame. He wrote in the *Chronicle* article:

> Blogs and prestigious university appointments do not mix terribly well. That is because top departments are profoundly risk-averse when it comes to senior hires....
>
> The trouble with blogs is that they seem designed to provoke easy doubts. Blogs are an outlet for unexpurgated, unreviewed, and occasionally unprofessional musings. What makes them worth reading can also make them prone to error. Any honest scholar-blogger—myself included—could acknowledge a post or two that they would like to have back....
>
> There are other risks. At Chicago, I found that some of my colleagues overestimated the time and effort I put into my blog—which led them to overestimate lost opportunities for scholarship.... Today's senior faculty members look at blogs the way a previous generation of academics looked at television—as a guilty, tawdry pleasure that should not be talked about in respectable circles.
>
> In some ways, this problem is merely the latest manifestation of what happens when professors try to become public intellectuals. Most members of the academy unconsciously accept the maxim that "foolish names and foolish faces often appear in familiar places." Blogging multiplies the problem a thousandfold, creating new pathways to public recognition beyond the control of traditional academic gatekeepers or even op-ed editors. Any usurpation of scholarly authority is bound to upset those who benefit the most from the status quo.

Blogging pseudonymously is one remedy to such vulnerabilities, allowing you to be as controversial and provocative as you like. Pseudonyms are accepted in the blog culture, although you must weigh the benefits of anonymity against the possible costs of being unmasked. A pseudonymous blog also means you will not be as widely quoted, and your influence will be less.

Why to Blog

Despite these cautions, however, there are also excellent professional reasons to launch a blog, particularly one covering a scientific topic:

- **A blog can give you instant feedback on your ideas, which can be invaluable in keeping your research and thinking on course.**
- **A blog can raise your profile in a field.** Whether you write a blog alone or moderate it, your blog presents you as an authority who has at heart the interests of your field. Your blogging also will put your opinion into the "Googlesphere." That is, people who do a Google search on that topic will

find your blog posts on the results list, ensuring that your voice is heard on the topic. A blog also makes you a member of the "media," giving you entrée into conferences and the attention of companies interested in exposure on your blog.

- **A blog can help knit a scientific community together, encouraging cohesiveness and constituting the equivalent of an extended virtual journal club.**
- **Blog discussion can document current thinking on a topic and can offer social and scientific context beyond that given by review papers.**
- **A blog can document a discussion.** While a verbal chat is evanescent, a blog discussion preserves ideas because the exchange is automatically documented as searchable text. This combination of interactivity and documentation can provide a fertile ground for developing new ideas.
- **A blog tends to encourage people to enter into a discussion who might be shy about expressing opinions in a seminar or class.** Teachers who use blogs for their classes report a substantive give and take that includes more students than does classroom discussion.
- **A blog can broaden the community of colleagues with which you interact, perhaps attracting people beyond your immediate area who can offer a new viewpoint and new ideas.** Economist J. Bradford DeLong has dubbed his blog "an invisible college, of more people to talk to, pointing me to more interesting things." He declared in an essay in the *Chronicle of Higher Education* that "My invisible college is paradise squared, for an academic at least."

More broadly, blogs that recount the personal experiences of doing research also offer a humanized, and interesting, view of your profession. They reveal the working of the gears of science—although they may grind noisily during the usual scientific debate—showing how lively, engaging, and fun research can be, says Johns Hopkins public information officer Joann Rodgers: "We have such stereotyped views of scientists—that they like to work alone, they like to work in their head. But science today is not like that; it is a very collaborative enterprise. And while there are scientists who fit that profile, there are also those who are artists, activists, wonderful conversationalists, good writers. For some, blogging might be a very good outlet."

A popular blog also helps fight pseudoscience and enriches the stream of scientific information to the public. For example, the RealClimate blog advertises itself as "climate science from climate scientists." The blog's description says it aims to "provide the context for climate-related news stories that is often missing in the mainstream media and to explain the basics of our field to the often confused, but curious, members of the public. In particular, it has provided rapid reaction to misuses and abuses of scientific results by policy advocates across the spectrum."

Blogging can also benefit your communication ability by jarring you out of the comfortable, technical writing voice of the researcher, giving you experience at a more accessible writing style. Bora Zivkovic of the Public Library of Science says he can tell when a scientist with blogging experience posts a comment on a scientific paper. "If they like the paper, they say 'I like the paper,'" says Zivkovic. "They don't like the paper, they say 'I don't like the paper because this, this, and this are wrong.' Very blunt, very straightforward, very simple English anybody can understand. But when you have people who are scientists with no experience with blogs, they tend to post comments that go on for an entire paragraph with 'Congratulations my dear colleague on this paper'—being very nice and very diplomatic, and then proceeding to destroy the paper point by point." A blog can also give you perspective on your own work, perhaps reminding you of what motivated you to get into research. The chance to step away from the lab bench and write essays on your topic can be a refreshing change from worrying about the experiment of the day.

Despite these uses of blogs to communicate, even the most well-done blog will not take the place of balanced, professional media coverage of your topic. For example, comments Juliet Eilperin of the *Washington Post* about RealClimate, "I definitely think RealClimate is very useful, and I think it is a good resource for journalists, but I don't think you should be lulled into thinking that is the way you can reach the public directly and bypass the media who still are best equipped to translate your scientific information."

As a final argument in favor of blogging, DeLong points out that blogging at universities could also be considered an integral part of some key academic missions:

> A lot of a university's long-run success depends on attracting good undergraduates. Undergraduates and their parents are profoundly influenced by the public face of the university. And these days, a thoughtful, intelligent, well-informed Web logger... is an important part of a university's public face.
>
> A great university has faculty members who do a great many things [including] the turbocharging of the public sphere of information and debate that is a principal reason that governments finance and donors give to universities. Web logs may well be becoming an important part of that last university mission.

Indeed, some universities formally recognize blogging's value in contributing to their "turbocharging" mission, by operating blogs that offer the public access to their faculty's knowledge and insight. One excellent example is the Science Life blog of the University of Illinois Medical Center, which offers information on clinical and basic medical advances.

Blogging Successfully

In developing a blog, first choose a topic that warrants an extended public dialog. Make the topic broad enough to give room for useful discussion. For example, a blog on a single enzyme is too narrow, a blog on an entire organism too broad, but a blog on a biological signaling pathway might be just right. If you plan to blog collaboratively, choose participants who have a shared sense of purpose, social ties to one another, and mutual trust that all will blog responsibly.

Once you have decided on the topic and participants, here are guidelines that will ensure your blog looks professional and that it is well written and broadly read:

- **Establish a brand for your blog by getting a specific URL for your blog, rather than using the URL supplied by your blogging service.** For example, I use WordPress for my blog, but rather than using the WordPress-supplied URL, my blog uses the URL *ResearchExplainer.com*.
- **Consider hosting the blog on your own Web site, rather than having it hosted on the blogging software site.** Thus, each blog post is registered by search engines as new content on your site, enhancing your search engine ranking.
- **Have a professionally created banner logo and "favicon," the small symbol that appears next to your blog URL.** Also, consider customizing the template, although the templates that blogging sites supply are perfectly serviceable.
- **Practice first.** Set up a temporary blog on such sites as Blogger.com, and write blog posts for a while without making them public, or post entries under a pseudonym, until you feel you are ready and can manage a blog.
- **Develop categories that reflect your target audience's interests.**
- **Include an RSS feed.** Such a feed automatically sends your blog posts to readers who subscribe to the feed. Most online blogging services have RSS syndication built in to them.
- **Adopt a casual blogging "voice."** Write more like you are talking to your colleagues over coffee than like an omniscient purveyor of scientific truths. This casual voice is necessary, since otherwise you risk coming across as a pompous know-it-all, rather than your friendly neighborhood researcher-blogger. Also, take care to limit jargon if the blog is meant for a broader public.

- **Post comments on other blogs, with a link back to your blog.** Commenting gives you experience using the blogging voice and also advertises your blog. Comments should be informational, not promotional.
- **Make your headlines concise and compelling.** They should tell what the blog post is about, attract reader interest, and promise a benefit such as new information.
- **Extensively tag your posts with key words, so that searches on those key words will bring up your blog.**
- **Make your entries substantive, perhaps a couple of hundred words, not just a few sentences.**
- **Voice an opinion rather than just conveying facts.**
- **As indicated above, be willing to post at least weekly over a long period of time.** Launching a blog and letting it wither is not good for your public persona.
- **Before you begin to advertise your blog, produce as many as a month's worth of entries, so that when people arrive, they will find significant content to read and respond to.**
- **Ask friends and colleagues to review initial blog posts for content and approach.** Is the tone right? Are you too heavy-handed or too timid in your opinions?
- **Make your blog a valuable information resource.** Report on new developments and products, review new books and papers, discuss hot issues, and highlight what other blogs are saying.
- **Go beyond text and include images and embedded videos.** These are easily gleaned from YouTube and other sources (see the online resources for chapter 3 for a list).
- **Use Google Alerts to monitor media reportage on your topic and cite those stories in your blog.**
- **Invite guest bloggers to write for your blog, and interview experts in your field.**
- **Invite comments and respond promptly and positively to those who comment.** Active commenters also offer a source of regular contributors.
- **Periodically "poke the bear."** Launch new discussions and conduct reader surveys and polls to elicit opinions. The online reference section includes links to survey systems such as SurveyMonkey and PollDaddy and tips on using them.
- **Use your blog to "crowdsource."** If you are writing an article or giving a talk, post a draft and invite comment. For example, Chris Anderson, author of the 2006 book *The Long Tail*, which postulates that the

Internet has altered the mass marketing of products, developed the book quite publicly by posting chapters and inviting comment on his blog *LongTail.com*.

- **Network with other blogs.** Add a "blogroll"—a list of other blogs—on your blog, and include links to other blogs in your posts. Consider participating in a "blog carnival," a magazine-like collection of blog posts on a particular topic.
- **Market your blog.** Submit it to such listings as Google Blog Search, Technorati, and BlogPulse.
- **Get yourself on media blogs by posting on your blog a comment on a media story.** That media blog will then list your blog as the source of a comment, and its readers will learn of your blog. Also, posts on your blog will be highlighted on such aggregation sites as Blogrunner, which automatically monitors news articles and blog posts and posts them on its site and distributes them via RSS feeds.
- **Add social bookmarking buttons to your blog posts, so people can tag them on those sites.** Social bookmarking is discussed later.
- **Monitor the statistics on your blog to determine where traffic is coming from.** That traffic flow can yield tips on how to reach those audiences more effectively.

The technical requirements of blogging are relatively straightforward. You can easily create a blog using popular blogging software such as WordPress. And blog-tracking by Technorati will index your blog among the 110 million or so that it tracks.

Tweet Your Research with Microblogging

So-called microblogging services, most notably Twitter, enable you to broadcast short messages, dubbed "tweets," instantly to people who have signed up to receive them. The messages can appear on the services' Web sites or be received via e-mail, your Web browser, mobile phones, or instant message systems. The default is for your messages to be public, but you can restrict them to a specific group. Answers to tweets can also be public or directed privately to a particular user. Or, by sending a public tweet to a user designated as @personsname, you can make it clear that the answer is meant for a specific person. Also, Twitter allows message threads to be tagged so you can distinguish a specific conversation from the general flow of messages. Besides having a personal Twitter account, you can create accounts around a topic, group, or event. You can search

Twitter for key words about you or your work using Twitter search, or by monitoring Twitter using Twilert or TweetGrid. These services automatically update you about tweets on a specific topic.

Besides Twitter, other microblogging services such as Plurk and Tumblr, offer additional features, including transmission of photos, links, and video. Also, the social networking sites discussed below, such as Facebook, offer quick update features similar to these microblogs.

It might seem that such short messages—in the case of Twitter, only 140 characters and spaces—might render microblogging all but useless for explaining research. However, such brief instant communications can have their place in your communication strategy. The power of microblogging lies in its immediacy and interactivity. For example, you might use microblogging to keep in touch with your research team as they go about their daily business. You and your colleagues can use Twitter to "crowdsource," soliciting help with experiments, alerting each other to events, asking questions, or taking informal polls. During symposia, microblogging enables easy coordination of activities and sharing of information on speakers, posters, and discussions with colleagues. During expeditions or field trips, frequent tweets can give your colleagues, as well as general audiences, useful updates and a sense of being there. If you are an educator, you could tweet your students on new assignments or conduct a discussion.

Microblogging can play a useful role in communicating with broader audiences, because its messages can create a kind of "virtual intimacy" with people who read your tweets—especially given the informal, chatty Twitter style. Twitterers often not only share useful information about their interests but also offer personal notes, hence the trademark Twitter question: "What are you doing?"

While such intimacy might seem frivolous, it can cement personal bonds—both within professional groups and with key lay audiences from students to donors. These personal bonds can give people a positive, friendly image of you and lubricate communications, making it easier to attract audiences to the more substantive information on your Web site and in your news releases.

For example, the clever communicators at NASA's Jet Propulsion Laboratory used Twitter to involve people in the progress of its Phoenix Mars lander. During the mission, a JPL communicator broadcast short, informal messages about progress of the Phoenix Mars Lander. The tweets, which pretended to be first-person messages from the lander, included such declarations as "Atmospheric entry has started. Time to get REALLY nervous. Now I'm in the 'seven minutes of terror.'" Many other NASA projects are now using Twitter as a channel to reach their audiences.

The many Twitter-related services for enhancing and managing Twitter feeds are listed in the online resources for this chapter. A few tips to be a good Twitterer:

- **List full information about yourself on your Twitter page, including your photo, Web site URL, and blog link.**
- **Promote your Twitter account in your e-mail signature line and on your blog and Web site.**
- **Use a conversational rather than a voice-of-God tone.**
- **Be open and honest.** Be frank about problems, frustrations—although remember that your tweets are public.
- **Share useful information with your audiences, for example, a tip, idea, or product review.**
- **Advertise events, including your talks and others of interest.**
- **Twitter often, even several times a day.**
- **Ask for ideas and input to spark discussion.**
- **Follow many other Twitterers, as well as the people who follow you, to learn of their activities, interests, and problems.**
- **Include URLs in your tweets, shortening them using SnipURL, TinyURL, or another URL-reducing service.**

Be a Podcaster

By producing podcasts—downloadable audio segments in an mp3 format—you can reach people who prefer to listen to your research story, for example, while driving or exercising. Podcasting is also more intimate and personality oriented, giving your audiences a feel for you as well as your topic. Of course, podcasts are especially effective if your work has an audio component—from bird songs to the pocketa-pocketa of a research machine you are using. Beyond disseminating information, podcasting also gives you experience in producing succinct, engaging explanations of your work and your field.

To sample some podcasts, explore the Apple iTunes Podcasting site. It includes not only a vast collection of podcasts to which you can subscribe, but also tutorials on creating and publishing podcasts. Apple also operates iTunes U, which distributes university educational content such as lectures. Other concise, interesting podcasts are Astronomy Cast, *Nature*'s podcasts, the *New York Times* science podcasts, and *Scientific American*'s "60-Second Science."

Particularly interesting are the podcasts of *Technology Review*, which produces podcasts of its articles using the text-to-speech system AudioDizer. The system's artificial voices—while not as alluring as the throaty purr of Lauren Bacall or the resonant baritone of James Earle Jones—are surprisingly understandable. Also, while AudioDizer will probably not accurately recite your technical papers, it is perfectly adequate for popular articles and general text.

An excellent guide to podcasting is the "How to Podcast" Web site. See the online resources for this chapter at *ExplainingResearch.com* for links to this and other sites discussed in this section.

Generally, your podcast should be short, concise, and personable, no more than 10 to 15 minutes. Use an expressive, enthusiastic "radio voice," which you can develop by listening to yourself on tape. Do you come across as bored or excited? Is your voice a monotone or interestingly modulated? In producing your podcast, you can be extemporaneous or script it, using the same process as writing a video script. To make your podcast more interesting, you can add relevant sounds and can interview experts, either in person or over the phone.

Steve Mirsky, producer of *Scientific American*'s podcasts, recommends investing in high-quality equipment to produce a professional-quality podcast. Total investment for such equipment is about $1,000. For example, Mirsky uses a Zoom H4 Digital Handy Recorder for recording face-to-face interviews. Although the recorder has good built-in microphones, he adds Electro-Voice brand RE50/B professional microphones, attached to the recorder with high-quality XLR cables. For recording over the phone, he uses JK Audio Broadcast Host Digital Hybrid, which is a component that enables high-quality, separate recording of host and caller voices.

To produce his podcast, he uses MAGIX Audio Cleaning Lab software to remove the hiss and hum from audio files, and Sony Sound Forge Audio Studio to capture a phone feed directly into the computer and to edit it. Such editing enables removal of noises, long pauses, and verbal glitches. For adding musical effects, Mirsky uses the Sony ACID Music Studio. Other editing software includes Wavepad, Adobe Soundbooth, and Apple Soundtrack Pro. If you do not wish to invest in such equipment and software, your institution's news office or media center might offer equipment, studio time, and perhaps editing services.

In distributing your podcast, post it on your blog and Web site, which should include an archive page with all your podcasts. You can also compile your podcasts on CD and offer them on your site. The commercial services Audio-Acrobat and Liberated Syndication offer a range of audio and video Web services, including recording and syndication. You can also syndicate your podcasts using the iTunes store, Podcast.com and/or FeedBurner. And, you can have your podcasts featured on the NSF science news site Science360, even if your work is not NSF-sponsored. If you are willing to commit to a long-form format, you can host a live, call-in program on BlogTalkRadio, which is automatically archived and made available as a podcast on iTunes.

Become a Social Networker

Popular social networking sites such as Facebook, MySpace, and LinkedIn can also be surprisingly useful to communicate your research. A personal page on such a site, especially Facebook and LinkedIn, renders you and your work more visible on the Web, especially since you can link back to your Web site.

Some journals, such as the *Public Library of Science*, have Facebook groups aimed at fostering scientific discussions. Also, BioMed Central includes a link to Facebook in each journal article, enabling posting of comments on the article. Facebook also hosts a wide range of science-oriented groups. Although LinkedIn is a more business-oriented networking site, you might find it useful—especially if your research has the possibility of yielding commercial applications. Some social networking sights enable you to form online communities specific to your interests. For example, FriendFeed enables users to form "rooms" specific to a topic or event, on which they can share and discuss Web pages, videos, photos, and other content. Using Flickr, you can share photos with a public or private group, and using YouTube, you can create a branded "channel"—a customized page that includes your profile information, videos, favorites, and other content.

More germane to your interests as a researcher are social networking sites for scientists, such as *Nature* Network, ResearchGATE, or SciLink. The systems enable you to form groups for your laboratory, department, or institution, as well as topic-based groups. The online resource section includes a list of such services, as well as the blog SciTechNet, which covers social networking services in science and technology. As with Twitter, while the professional aspects of such sites are obviously useful, the social features might seem a bit frivolous. However, they do have value in encouraging cohesiveness and collaboration. Social networking helps you "read" your fellow scientists. Thus, in some ways, online social networking sites constitute an extension into cyberspace of the traditional face-to-face networking, long a part of the scientific culture. Certainly, the local coffeehouse and/or tavern have been the site of more than a few scientific brainstorms.

Capture Eyeballs with Social Bookmarking

"Social bookmarking" sites are another useful way to disseminate information on your research. Such sites as Digg, deli.cio.us, Mixx, Reddit, StumbleUpon, and Yahoo! Buzz enable users to mark articles that the site's users will be interested in. You can use these sites to bookmark content on your own site, entries on your blog, and articles about your work. You can also add Digg, del.icio.us, and other bookmarking buttons on pages of your site, so others can bookmark it. And, of course, you can monitor such services as StumbleUpon for articles about your topic, to post on your blog or Web site.

The social information service Squidoo enables you to establish a "lens" on a specific topic and to post material that relates to that topic. You can establish your research area as a topic and link to your Web site, as well as other content that will draw people to the topic. Thus, Squidoo will harness traffic already coming to Squidoo and redirect that traffic to your Web site.

Gather Online with Web Meetings and Webinars

Given that many of your audiences are technically sophisticated, consider reaching them using Web meetings and webinars. Web meetings enable you to conference with individuals and small groups. The Web meeting interface supports two-way audio and video conferencing, as well as the ability to present slides, Web sites, video, software, and documents and to write on a shared whiteboard. A webinar offers the same basic capabilities but enables presentations by multiple speakers to large audiences.

The simplest webinar is the "hybrid system," in which participants call in via telephone to listen to the audio, while they view slides on a special Web site. A more advanced webinar system features online audio to accompany the Web-delivered slides. Also, for participants with a slow Internet connection, slides can be downloaded ahead of time.

Web meeting and webinar software—including Adobe Acrobat Connect, DimDim, GoToMeeting, GotoWebinar, Microsoft Office Live Meeting, On24, Netviewer, and WebEx Meeting Center—are relatively inexpensive and can be mastered with a little effort. Also, Web video is now relatively simple and easy to use. One such Web video program is SightSpeed. Many of these services' Web sites offer demonstrations and free trials of their software. For examples of science webinars, see the *Science* magazine webinar series. Also, for online training on Web meetings and software, see vendor sites such as WebEx University.

Holding an Effective Web Meeting

To learn basic Web meeting techniques, start with smaller scale meetings before undertaking a webinar with a large audience. Other tips:

- **Use the minimum technology necessary to communicate for your first Web meetings.** For example, start with just a phone teleconference, in which people can see the material discussed on a Web site. As you become comfortable with the basics, graduate to more advanced features such as Web video, a whiteboard, and software demonstrations.
- **Keep Web meetings short: no more than 60 minutes.**
- **Keep slides and other visuals relatively simple.** Complex slides will be hard for attendees to digest, and large images might take time to download.
- **Practice displaying slides and other visuals you will use in the meeting.**
- **Distribute an agenda that clearly defines the meeting and topics.**
- **Close the door to your office and ask not to be disturbed.**
- **Use a wired land line not a cell phone for the meeting.** And do not use a speakerphone—use a wired headset or, if that is not available, a good handset.
- **Turn off cell phones or any other devices that might make a sound.**
- **If you use video, wear solid colors, rather than patterns that may be distracting.**
- **Close unnecessary programs on your computer to improve Web conference software performance.**
- **Go to the bathroom before you start, and have water available.**
- **Start on time.**
- **Introduce the agenda at the beginning—both orally and visually—and announce each point as it is reached.**
- **During the meeting, do not eat, drink, shuffle papers, check your e-mail, or do any other distracting activities.**
- **Keep the meeting moving and focused on the agenda.** Do not let participants ramble or present extraneous slides.
- **Engage the attendees by posing useful questions and asking for feedback during the meeting.**
- **Use good video technique for video Web meetings.** For example, make sure the Web camera is adjusted properly, focused, and centered and

that you are properly illuminated. Make eye contact with the camera, speak clearly, and do not fidget. Make no quick movements because they produce a distracting blur.

- **Wrap up the meeting by summarizing the results.**

Managing a Successful Webinar

Consider creating a large-scale webinar when you can justify it by the size of the potential audience. Webinars can be very effective for departments, centers, or schools to explain a broader topic or research program. Some tips on developing a webinar:

- **Carefully plan the webinar.** These plans should include establishing a goal, creating a compelling topic, recruiting effective, interesting speakers, developing quality visuals, marketing broadly, and creating an easy registration process with confirming e-mails.
- **Consider hiring a professional online moderator who understands how to conduct a polished event.** For example, such a moderator knows how to effectively handle transitions between speakers, Q&As, and polling.
- **Include plans for follow-through and feedback to assess the webinar's effectiveness and whether participants have additional questions or suggestions.**
- **Decide on the Web conferencing system, if possible with the help of professional techs in your institution.** Also, have a technical support person available for the webinar.
- **Schedule the webinar so that the most people can attend.** For example, hold a national webinar in midafternoon to be convenient for all time zones, and at midweek, since fewer people are available before and after weekends.
- **Promote the webinar.** Mail brochures early, and advertise on your Web site, in e-newsletters, and via word-of-mouth networks, asking people to tell others.
- **Create a special "landing page" for the webinar.** The Web page should offer preparatory background material on the topic and information on the technical requirements for the webinar. The page also should feature registration and a survey, so that you can obtain information on participants, including their level of understanding of the topic.

- **Send out a reminder to all registrants a few days before the event.**
- **Conduct practice sessions.** Refine the presentations, proofread materials, check the technology, and coach speakers on the particular demands of presenting on the Web. For example, they might consider using interactive polling and Q&A and chat features to keep the audience engaged.
- **Make sure presentations are concise and all material is relevant to the audience.**
- **Plan for catastrophes, such as loss of a presenter's connection or poor-quality voice transmission.**
- **Start the event precisely on time; even a short delay can frustrate an audience.**
- **Post a professional-looking welcome slide that lets audience members know they are in the right place.**
- **Have the moderator go over ground rules and how to use chat and polling features.**
- **Make sure all speakers use wired land lines and headsets, not speakerphones.**
- **Make sure all speakers are in quiet rooms to eliminate distracting background noise.**
- **Make sure speakers have a hard copy of their presentation, should there be a glitch in visuals.**
- **Allow plenty of time for answering questions that participants will type in using the webinar software.**
- **Keep participants until the end by promising an important piece of information at the webinar conclusion.**
- **Respond to all questions by e-mail or phone within 24 hours.**
- **Archive your webinars on your Web site. If you have multiple webinars, create a special page listing them.**

Integrate Your Social Media

Remembering your strategy of synergy, you can integrate the content on your Web site, blog, podcast, Facebook page, Flickr page, YouTube channel, and other social media sites, so that posting to one will feed content to others. The simplest such integration is to link the sites—for example, featuring a link to your blog on your Web site. You can also use the many plug-ins available on social media sites, such as WordPress to integrate sites.

Also useful are syndication sites such as FeedBurner and *Ping.fm*. FeedBurner automatically feed your content to multiple applications—for example, feeding your blog posts to the notes section of your Facebook page. More broadly, FeedBurner can enable you to automatically distribute your content to subscribers, who can receive it via a Web portal, news reader, or e-mail message. This content can include text, images, audio, and video. The FeedBurner site includes a comprehensive help section covering such integration and syndication.

Ping.fm enables a single posting to feed multiple social networks. For photos, you can post a Flickr "badge"—a bit of code—on your Web site that will automatically pull random Flickr images onto your site. And you can use the YouTube channel feature to create a video gallery that will embed videos into your Web site, blog, or Facebook page.

16

Write Popular Articles, Op-Eds, and Essays

Writing lay-level articles—whether a magazine article for *American Scientist* or *Scientific American* or a newspaper op-ed—can be a valuable way to explain your research and its implications. However, successfully writing popular articles requires understanding the considerable differences between lay-level and the professional writing you are used to. Certainly, the same basic rules of good writing discussed in chapter 6 apply. However, the style of lay-level writing is quite different than that of a professional article. This chapter aims to help you learn that style and navigate the sometimes frustrating process of lay-level publishing.

Why bother to write lay-level articles? Of course, an article for *Scientific American* or *American Scientist* offers professional advantages. Their readers will include researchers and students who will gain useful insight into your work and how it fits into the field. What's more, says former *American Scientist* editor Rosalind Reid, such articles reach across disciplinary boundaries, offering professional advantages:

> Our articles are written for a general educated audience, but that also means they reach scientists outside the author's field, because both kinds of writing are the same. So, these articles allow scientists to see connections with other fields that they wouldn't otherwise see because

> the language is so different. This causes real collaborations to happen; causes real joint funding proposals to happen; causes real students to come to you wanting to study your work who might come from a different field.

Lay-level articles in commercial magazines and newspapers portray you to the public as an authority in your field. Because such articles are, in essence, peer-reviewed by the publications' editors, they give you the same credibility among the public that scientific articles give you among your peers.

What's more, popular articles portray you as civic-minded—committed to sharing knowledge in your field with a broader audience. And, op-eds and essays portray you as a researcher concerned about the broader political and social implications of your field. Popular articles also communicate your work and your field with a sense of immediacy and intimacy that no scientific paper can offer. Thus, they can engage important audiences—students, administrators, donors, and your family—in a way that even the best-written technical description of your work cannot.

Lay-level writing also teaches you communication skills that will pay off in all your writing, including your professional writing. In making the editorial exertions to reach a lay audience, you will develop writing "muscles" different than the ones you use for scientific articles. Communicator Chris Brodie says of his transition from neurobiologist to writer/editor, "Now I see things in pieces of writing that I was blind to before. I start noticing its structure, and noticing really skillful turns of phrase. It is like going from being able to appreciate music to starting to learn how to make music yourself, and it is a whole new level of virtuosity."

Prepare Yourself to Write

Before you tackle the challenge of lay-level writing, decide whether you are willing to commit yourself to its rigors. And they *are* rigors. As sportswriter Red Smith said, "Writing is easy. All you do is sit staring at the blank sheet of paper until the drops of blood form on your forehead." The quote comes from *For Writers Only* by Sophy Burnham, a gem of a book on the pains and joys of the writing life.

Also, consider whether you have something to say that people will be interested in reading. Fortunately, your knowledge of your field gives you an advantage in that you can write with authority. However, successful lay-level writing means not only having special knowledge, but also figuring out what aspects of that knowledge will interest lay audiences. To gauge lay interest in your field, start by talking to your family and friends not in your line of work about your research. What interests them about it? What does not? For example, if you do theoretical astrophysics, they will probably be fascinated about the exotic physics of black holes, but not about your latest theories of the dynamics of accretion disks. Or, if you do molecular biology, they would be interested in how mitochondrial malfunction can cause disease, but not about the role of mitochondrial Bcl-2 proteins in apoptosis.

To write engrossing lay-level articles, you must also commit to learning about science journalism. One excellent source is *A Field Guide for Science Writers*, a collection of essays by science writers and editors that explores science writing techniques, markets, and genres. An excellent book on the craft of science writing is *Ideas into Words* by Elise Hancock. And a standard comprehensive journalism textbook is *Reporting for the Media.* Links to these and more sources can be found in the online resources for this chapter at *ExplainingResearch.com.*

Also, by reading the best lay-level articles and books about your field, you can learn how good science writing is done. These include news and feature articles in such publications as the *New York Times*, the *Washington Post*, *American Scientist*, *Discover*, *Science News*, *Scientific American*, and *Popular Science*. Also, the *Best American Science Writing* book series offers an excellent collection of articles. To see how the science writing "sausage" is made, regularly read the online Columbia Journalism Review Observatory, which offers analysis and critiques of science coverage.

In your reading, pay attention not only to the substance of the articles but also to how professional writers use anecdotes, explanations, quotes, and other elements in their prose. Also, pay attention to the distinctive structure and style of each publication's articles. For example, a typical newspaper feature story begins with an anecdote about a person or event, aimed at engaging readers. The article then goes on to explain the substance of the topic, using the person or event as an illustration.

Also read articles and books about how to write well. Sources such as the *Writer's Digest* magazine, books, and Web site can be very helpful. Consider taking journalism or writing courses and workshops, especially if you plan to write regularly at a lay level. Most schools and universities offer general journalism courses as part of their curriculum or as extension courses. Also, many larger universities offer specific courses in science and medical writing. A comprehensive directory of science writing courses and programs, developed by University of Wisconsin's Sharon Dunwoody, can be found in the online resource section for this chapter. Participating in writing workshops can be especially helpful because you will get real-time feedback on your work from other writers.

And you can dragoon friends, family, and colleagues into being the first editors of your works-in-progress, says Brodie. "Show your work in its formative stages to as many people as you can," he advises. "Not just the people in the lab, but somebody in a department across campus. See if they can follow your writing, and get them to tell you specifically where the problems are."

If you plan to do a lot of lay-level writing, consider joining such organizations as the American Medical Writers Association, the Authors Guild, the National Association of Science Writers, and/or the Society of Environmental Journalists. They offer background information, conferences, workshops, and an experienced network of writers who are quite willing to help you.

Be a Storyteller, Not an "Authority"

Just as the skills for lay-level writing differ from those for technical writing, so does the editorial attitude. In your professional writing, you assume the role of authority, but in lay-level writing you are primarily a storyteller. Your mission is to tell the story of your topic, and that includes telling the stories of the people and events behind it. "If you as a scientist are writing a story for a journalistic outlet, the reader expects you to be a transmitter, not a source, and you have to honor that socially constructed role," says Dunwoody, a professor of journalism and mass communication. "And in this setting if you, the scientist, takes the attitude that 'I am the expert, so I am now going to tell you the reader what you need to know,' readers will simply reject what you have to say."

As a storyteller, your job is to make your articles both informative *and* interesting. For example, you will be expected to spin anecdotes about your topic that entertain lay readers and draw them into the article. What's more, when you lure lay readers into your story, you lure your colleagues, as well. Your fellow researchers will also respond to such engaging elements as anecdotes in both your lay and technical communications.

Dunwoody, for example cites a study in which communication researcher Alan Hunsaker wrote a science story in two ways—as a complicated technical explanation and as a simpler more accessible story. He then measured how much information PhD-level readers versus lay-level readers learned from the two versions and how much they enjoyed them. As might be expected, the PhD-level readers got the most information from the technical version. But surprisingly, they enjoyed the simple version the most. "What that finding says is if you can tell a story, even a simple story, in a way that is readable and fun, your PhD colleagues will enjoy that story just as much," says Dunwoody.

Understand the Editorial Process

Journalistic writing, like scientific writing, has a standardized editorial process, and if you understand it, you are far more likely to be successful. Among the major components of that process are marketing and pitching articles, the editing process, and the publishing business.

Where to Market Your Article Idea

If you are just beginning to write lay-level articles, consider starting with local media, perhaps those in your own institution. All universities have alumni magazines, and some have research magazines that might publish an article on your research or your perspective on your topic.

The magazine of your professional association also might be interested in an article. These include *Bioscience*, published by the American Institute of Biological Sciences, and *Chemical & Engineering News*, published by the American Chemical Society. The online resource section lists links to author information Web pages for some of the major research and association magazines. You might also aspire to write for general science magazines, such as *Air & Space*, *American Scientist*, *New Scientist*, *Physics Today*, *Popular Science*, *Psychology Today*, *Science*, *Scientific American*, *Sky & Telescope*, and *Smithsonian*. The next section covers how to pitch articles to such magazines, and the online resource section includes links to their writing guidelines. For a broader list of publications and their editorial interests, subscribe to the *Writers Market* online resource. The site also contains a broad range of articles on writing and marketing.

If you have rights to your articles, you can also sell them online, using PayLoadz—a service that enables you to sell any downloadable content at a price you set. It might even be worthwhile to write popular articles specifically to be posted on your own Web site and sites that offer free "ezine" articles. Such ezine sites post articles that are available free to other Web sites, with attribution. These sites include *EzineArticles.com* and *ArticleCity.com*. Check out articles already posted, and match them in style, content, and length. Think of topics that fit into their existing categories. For example, for *ArticleCity.com*, if you are a psychologist, write an article that fits into the education or self-improvement category. If you are an engineer, write an article that fits into the gadgets and gizmos category.

Such articles need not be long or comprehensive, or take much time to write. For example, any astrophysicist can knock out a basic article on black holes without too much trouble, a soil scientist can summarize interesting facts about plain dirt, and a pediatrician can list the advantages of childhood immunization.

Posting such articles on your own site will lead to links to the site from others. And such links will enhance your search engine ranking. You might be surprised how such lay-level articles, distributed widely, can help your credibility, authority, and reputation as a civic-minded scientist.

Making the Pitch

In pitching an editor on a feature article you want to write, your first contact is usually in the form of a pitch letter. This is a sales piece to convince the editor that your topic will interest the publication's audience, and that you are just the person to write about it. So, the pitch letter must not only feature an engaging description of the topic and how you plan to cover it, but must also constitute a mini-showcase for your writing ability.

You can consult the expert advice section of *WritersMarket.com* for articles on producing successful query letters, as well as the archives of *Writers Digest* at *WritersDigest.com*. Some tips on writing a compelling query letter:

- **Know what the publication covers.** Read back issues, so you will understand the topics covered and the style of its articles. Editors will immediately reject a query letter from a writer who misses the editorial mark, in terms of either topic or style.
- **Know the publication's writing guidelines.** Most magazines post writing guidelines on their Web sites, and as mentioned previously the online resource section includes guidelines for major popular science magazines. Even if you do not plan to write for these magazines, review their guidelines to get a better sense of what magazine editors look for.
- **Know the editor.** Address the letter to a specific editor by name who you know is the right one. If you are unsure who to contact, check the Web site, or phone or e-mail the publication.
- **Hook the editor.** State your topic immediately and compellingly and why it is interesting and important. Do not "back into" the query by trying to be coy or cute. The hook can consist of a reader's problem the article will solve, a fascinating fact the article will explore, an intriguing question it will answer, or a dramatic personal experience the article will relate.
- **Be concrete.** Offer an anecdote that vividly illustrates what you will bring to the topic. Explain the angle you would take, who you would interview, and what they could say.
- **Know where your article could fit in the publication.** Cite the specific section of a magazine, newspaper, or Web site at which your article would aim.

- **Know what else has been written on the topic.** Your query letter should persuade the editor that your article will be fresh and unique, either in its topic or in your approach. So, indicate that you have surveyed other writing on the topic and know how your proposed article fits.
- **Keep it short.** Editors are busy. Keep the query letter to no more than 600 words or so. Do not include unnecessary personal history, flattery, apologies for your inexperience, or other subjective information.
- **Do not request advice or criticism.** Again, editors are busy.
- **Proofread it!** Make sure there are no spelling or grammar errors.
- **Describe your writing experience and credentials.** List other lay-level articles you have written and your qualifications for writing about the topic.
- **Propose visual possibilities.** Describe possible photos, artwork, or diagrams. Also, keep in mind that print publications have Web sites that can use video and animation.
- **Indicate if you are willing to write on spec.** If your writing experience is limited, you might have to offer to write the article "on spec," that is, without a contracted assignment. Spec writing is more feasible for short pieces than for longer pieces that will require considerable research and interviewing.

Keeping "Lazy, Dumb" Editors Happy

Once you get an assignment, as you write the article remember that editors are a "lazy" bunch—at least, they might seem so to an outsider like you. After all, they won't take the time to edit your rough draft or coach you on their writing style. Rather, they insist on having a draft article arrive in their inbox that exactly fits their publication's requirements. Of course, this "laziness" is no such thing. It reflects that editors are extremely busy professionals, especially those at large publications. And, they quite correctly see any deviation from their publication's requirements as evidence that you are not a serious writer. So, even though you might consider yourself a "creative" writer, do not be creative with the requirements of an article. Strictly observe word count, editorial conventions, and style, just as you would for a scientific paper. Editors also like articles to arrive along with ideas and sources for artwork and contact information for your sources. Make yourself a full-service writer and you are more likely to get assignments.

Also, the best editors are "dumb"—in a very smart way, wrote *Scientific American* editor Mariette DiChristina-Gerosa in *A Field Guide for Science Writers*:

> Let's be honest. Editors, as any writer will tell you, aren't all that bright. They may say they're looking for stories that will teach something

> important about the way the world works, but mostly they want to be entertained. They can't follow leaps of logic. They get distracted by elaborate prose, and they have no patience for boring, factual details. They get confused by too many characters in a narrative, or they're easily irritated by extraneous quotes. And they don't like big words very much, either.

"In other words," wrote DiChristina-Gerosa, "we editors are a lot like the readers that we—and you—are trying to reach." She declared that editors "live and breathe our readers' way of life." So, keep editors, and readers, happy by giving them entertaining, dynamic, tightly written prose.

Tolerating Good Editing

The editorial review process for lay-level writing can be just as frustrating as that for peer review of scientific articles. Just as you invariably believe you have submitted the perfect scientific paper, you will likely harbor the illusion that you have submitted the perfect draft article. However, be prepared to be edited to high journalistic standards. Lose your writer's ego. If you are lucky, an editor might have only minor nitpicking questions. But more likely, your article will undergo major dissection and reorganization—especially if you are writing for *American Scientist* or *Scientific American* and are not used to lay-level prose.

"There are layers of editors, and those editors are going to come back with a lot of questions and a lot of rewrites," warns former *Smithsonian* magazine editor Sally Maran. "There is almost nothing that doesn't get rewritten, and you can't let that ruin your ego. Even professional writers have to rewrite." Your surprise, perhaps even shock, at such rewriting might come because you do not appreciate the substantive role of the editor. "I remind authors that the editor works for the audience," says Reid. "The editor's job is to know the audience and to help you connect with that audience. However, writers are often under the misapprehension that they have just encountered a presumptuous editor whose job is normally to fix a misplaced comma."

Enduring Great Editing

If good editors are merely annoying, great editors can be a royal pain. The best editors will engage a piece in minute detail and with deep insight, and the result will likely be a blizzard of questions and editorial suggestions that might at first seem to be the work of an overly obsessive, even pathological personality.

However, if you sit in a dark room for a while, let your frustration dissipate, and look at the editorial response with an open mind, you may come to understand that it is both incisive and helpful. Those edits have come from an editor whose acute editorial eye sees with crystalline clarity flaws in your article's writing, fact, and logic.

If you are lucky—as I have been several times in my writing career—you will encounter a brilliant editor. As a novice writer, it might be difficult to get over your beginner's ego and learn to recognize and appreciate such expertise. However, once you have surmounted your ego, you will realize that the "pain" of dealing with a great editor is well worth it. The resulting article will be far better than you could have ever made it alone.

Understanding an Arbitrary Business

Beyond the pain of editing, the publication process bedevils writers with many other frustrations. For example, most article pitches will be rejected, given the limited space of magazines and the vast numbers of article proposals editors receive. Even the best professional writers often suffer many rejections, in order to sell an article to a top magazine. So, be prepared to pitch many articles to many publications before having one accepted.

Once an article is accepted, the editorial process can often be a crap shoot. Your article could be at the mercy of editors' personalities, the vagaries of the publishing schedule, and unforeseen developments in your topic. My own magazine articles have been victims of such random fates. Inept editors have mangled articles so badly that I have asked them to remove my byline. Other articles took an excruciatingly long time to make it into print—one taking four years!

The business of lay-level publishing can also be frustrating and arbitrary. For example, publishers may ask you to sign "all rights" or "work-for-hire" contracts, in which they reserve all rights for your writing. These contracts are fine if you fully understand their implications. However, remember that copyright law favors the writer. The instant you write something, it is copyrighted in your name. Usually, when you sell a piece to a publication, you are selling it one-time rights, after which the rights revert to you. Also, freelance writing pay is lousy. Many decades ago, the typical pay for a freelance science piece was $1 per word. Today, the pay is roughly the same. Keep your day job.

The best protection against the vagaries of the writing business—if you plan to do considerable freelance writing—is to join one of the professional writing associations listed previously. They can offer advice and expertise in negotiating contracts, understanding fees, and protecting your publishing rights. Now that you have been initiated into the editorial process, here are some guidelines on writing popular articles.

Write Op-Eds and Essays

Writing short opinion pieces and essays in your field is perhaps the simplest way to get started writing for the public. The op-ed pages of newspapers are always interested in expert opinion on issues of the day. Other ready outlets for essays on your field are your professional association magazine and others in your discipline. To dip your editorial toe in the journalistic opinion waters, consider writing letters to the editor at first. The time investment is minimal, and you will get experience working in a journalistic setting.

As discussed above in the section on pitching articles, prepare yourself by reading op-eds and essays in the newspaper or magazine you are targeting. Note the style, length, and subject matter. Also, as mentioned previously, do not be "creative," in the sense of going beyond that style, length, or subject matter.

Crafting Your Op-Ed

Here are some tips for writing a successful op-ed:

- **Be a relevant expert.** You are most credible as an op-ed writer if your expertise is closely related to the issue in the news. The fact that you are a molecular biologist will not make you a credible expert on a clinical medical issue, for example.
- **Have a fresh opinion.** Your op-ed must express an opinion—the fresher and more counterintuitive the better. If you are merely supporting conventional wisdom or agreeing with the story you cite, there is no reason to publish your op-ed. And if you are merely adding a bit of insight to the story, you should have written a letter to the editor.
- **Be willing to express that opinion in a concerted, even controversial way.** Researchers love to write pieces in which they are "two-handed"—that is, they give "on the one hand" and "on the other hand" discourses, and end up calling for more research. While such a stance might satisfy your scientists' sensibility, it will not interest an editor.

- **Put the opinion at the top.** Researchers love to preface their articles with background, but an op-ed should lead with the opinion that you are going to support.
- **Be specific.** Refer specifically to the newspaper article, issue, or event on which you are commenting.
- **Tell readers why they should care about your opinion.** Your op-ed should contain the equivalent of "While the article in the February 15 *Gazette* cited experts questioning the existence of the Easter Bunny, this misguided position will deprive millions of children of their rightful Easter baskets if it prevails."
- **Aim for a single, compelling point.** In a short op-ed, you cannot effectively discuss more than one aspect of an issue. For example, write "The Easter Bunny undoubtedly exists." Not "There is an extensive body of research on the biology and ecology of Easter bunnies, and here is a summary of that work."
- **Offer specific remedies for a problem.** Do not just rail against an injustice or inefficiency or, as indicated earlier, call for more study. Advocate for a specific law or remedy that will help correct it, for example, "Congress should enact laws already proposed that mandate minimum chocolate requirements for Easter baskets and their availability on Easter morning."
- **Use concrete examples and compelling statistics to support your argument.** For example, "More than 50 million children receive Easter baskets each year, producing one trillion child-joy units as measured by the U.S. Department of Children's Happiness."
- **Use active voice.** It is very much to be desired if your opinion is to be heeded by readers. See how boring that last sentence was?
- **Get personal.** Write in first person or brag on your own experience. Surely you recall your own delightful childhood experiences searching for your Easter baskets. Or, perhaps there was that horrible Easter when your brother ate most of your candy.
- **Summarize your argument in a bang-up conclusion.** Give readers a ringing declaration that they will remember: "Save the Easter Bunny from extinction, and we save the happiness of our children."
- **Avoid jargon.** Have a nonscientist read your draft to make sure you have written Easter Bunny rather than *Sylvilagus transitionalis.*
- **Strictly observe the 600–800 word length.** As discussed above, "lazy" editors delight in receiving articles that are just the right length, and 600–800 words is the widely accepted length for op-eds. If your piece is longer but accepted nonetheless, the editor will trim it down. Better that you submit a piece you crafted to the right length yourself.

Marketing Your Op-Ed

Here are tips to ensure that your op-ed sees publication:

- **Be timely.** News ages quickly. Start writing the minute you read a story, and have your op-ed ready to submit as quickly as possible.
- **Be prepared.** If a controversy in your area frequently makes news, have a basic op-ed prepared, so you can adapt it and respond quickly to a breaking story.
- **Work with your news office.** It may run an op-ed distribution service or have connections to editorial page editors. With a distribution service, the office can submit op-eds to subscribing newspapers. The office might even be willing to edit or ghostwrite your op-ed.
- **Aim "low."** Unless you have something stunningly compelling to say, and you are willing to face considerable odds, do not submit to the *New York Times* or the *Washington Post*. Those newspapers receive massive numbers of op-eds. Especially if you are just starting out, write for local or regional newspapers.
- **Extend your op-ed's reach.** Post it on your Web site, and share it with colleagues and administrators. It might lead to other opportunities, such as invitations to talk about your field or discuss your views with legislators or other opinion leaders.

Write Feature Articles

Writing long feature articles can be among the most satisfying lay-level writing. An article in *American Scientist* or *Scientific American* enables you to wax eloquent on your field and its future. And an article in *Smithsonian*, *Natural History*, *Discover*, or *Psychology Today* allows you to exercise your creative journalistic voice—a very satisfying process for a writer who might have felt straightjacketed by the constraints of scientific writing.

In writing a publishable feature article, pay attention to the basic writing guidelines discussed in chapter 6 and the news release guidelines in chapter 10. Besides these, there are some particular journalistic requirements for magazine articles.

Ditch Your Colleagues

While almost all your writing has aimed at your colleagues, lay-level writing emphatically does not, and you must adjust your thinking. "You just have to get your colleagues out of your head," says Reid. "That is the hardest thing to get across when I am editing someone. They are so anxious about how a particular colleague will read a piece if it is simplified, or doesn't use a particular piece of jargon or a particular equation."

To rid oneself of the censorious ghosts of one's colleagues, Reid advises researcher-writers to broaden their perspective.

> They can spend time off campus in the general culture to understand the difference between the outside world and the one in which they normally write and give papers. I also tell them to imagine what they would tell somebody in a bar sitting at the next barstool about their work. Or, I suggest that they show the piece to their spouse or somebody else in their family and see if they can honestly understand it. Or, I suggest they give a talk to a class of naive freshmen and record it or use their notes to make that into their article, because a good talk is a good article. Your job is the same in both cases—to keep the audience on the edge of their seats the whole time, to keep it interesting.

Know Your Readers

As chapter 1 emphasized, knowing your audience is critical to explaining your research, and this understanding is particularly key to writing engaging popular articles. Not only are popular and scientific audiences different from one another; the readers of popular magazines are different from one another, points out Sally Maran: "You cannot write a generic science story that could appear in *Smithsonian*, or *Popular Science*, or the *New Yorker*," she says. "Each one of these stories is going to be told in a different way. For example, *Smithsonian*'s approach to many of its articles is to tell a mystery story about a piece of science and to have fun with that mystery." Maran emphasizes that such mystery stories must appeal to a broad spectrum of readers. "Scientists don't come to *Smithsonian* to read science stories," she says. "I felt that historians came to us to read our science and art stories; and art history people come to us to read the science and history pieces."

Know What Your Readers Do Not Know

Be acutely aware of what your readers do not know about your subject. To sensitize yourself, besides mingling with nonscientists, consider the kind of exercise

Sharon Dunwoody gives science students in her course on communicating research:

> We ask the students to prepare an interactive science demonstration for a science day we have on campus. And we ask them to go out and explore what their audiences might know and think. They are often very surprised at the result.
>
> For example, one group was interested in doing a demonstration on carbon sequestration, and they had developed an idea of what they wanted to do. They went out to talk with people and came back horrified, because they said the people they talked with didn't even know what carbon was. It hadn't occurred to them that was an issue. So, they backed up a bit and incorporated into their demonstration what carbon is, why it is critical to life, and why we care about where we are storing it.

Organizing to Grab and Hold Readers

For almost a century, scientists have paid obeisance to the god of organizing papers IMRAD—Introduction, Methods, Results, and Discussion. In starting your lay-level article, you should renounce IMRAD in favor of a more enticing lead-in. While IMRAD is a logical way to engage scientific audiences, it does nothing to lure lay readers to read your articles. You must plunge them immediately into your story using one of the grabber beginnings listed below. "I tell authors that they must figure out how they are going to motivate the reader, either through self-interest or through curiosity with some fascinating story," says Reid. "They cannot depend on the same motivation as in journal articles; that the reader needs to keep up with the field."

And tighten your introductory text to make it as short and snappy as possible. A saying among writers is "Grab a reader with the first sentence, and you have them for a paragraph. Grab them with the paragraph, you have them for the page. Grab them with the page, you have them for the article."

There are many types of grabber beginnings, as illustrated by the examples from my articles listed in the online references and resources for this chapter at *ExplainingResearch.com*. These grabber ledes are those that

- Tell a story about someone involved in the article
- Tell a personal story
- Describe a researcher's subject
- Describe a researcher's exotic environment

- Describe vividly even the mundane environment of a laboratory
- Describe a moment of discovery, reconstructed from an interview
- Describe a vignette of field work
- Describe a treatment or other process
- Introduce a research facility

Beyond the grabber beginning, however, your article's overall organization is somewhat IMRAD-like—what Reid calls an "hourglass" shape: "An hourglass organization means leading with big thoughts, big ideas, important stories; then discussing details in the middle; and having broad ideas at the end—the future, research questions, and so on." However, unlike IMRAD, you do not end with mere navel-contemplating "discussion," but with a bang—by revealing the ending of an anecdote introduced in the beginning, giving a ringing quote, or summarizing the topic with a pithy conclusion.

Use Techniques of Fiction

Without sacrificing scientific accuracy of your articles, you can use techniques of fiction to get your ideas across in a way that engages audiences. The online resource section includes books on writing fiction that can teach you those techniques. They include two-time Pulitzer Prize–winner Jon Franklin's classic *Writing for Story*, and *The Writer's Digest Handbook of Novel Writing*. Here are some of the guidelines these books recommend:

- **Tell stories.** As emphasized previously, your prime role is that of a storyteller, so think of anecdotes that will illuminate and dramatize the topic of your article.
- **Get inside your characters' heads.** When you interview people, ask them what they were thinking, how they felt, and so on. Such details bring your story alive and lure readers in.
- **Be a camera.** Craft word pictures as if you were a camera. Frame the scene in your mind and describe what your mind's eye sees. Use such framing, along with mental panning and zooming, to create your description so that it serves your purpose of communicating your topic. For example, a scene might begin with the description of a cluttered laboratory late at night, then "zoom in" to a biologist bent over a petri dish, and from there zoom in to the bright red streaks of bacterial colonies on the translucent gray agar.
- **Describe characters.** Describing people's appearance or mannerisms might not seem to have anything to do with explaining concepts. However, such character descriptions draw readers into your topic. They

want to "see" that burly, balding scientist with a faded Semper Fi tattoo crouched over the field seismometer, his sweat droplets dotting the desert sand as he roundly curses the glitching instrument.

- **Show rather than tell.** Rather than simply telling—for example, that a scientist was surprised at a discovery—show that surprise: "Her eyes widened, and she gasped faintly and reached a finger out to gently touch the computer screen, as she saw the image of the opalescent spiral galaxy leap into view."
- **Write for the senses.** Describe scenes using as many of the five senses as possible. Describe not only what something looks like, but what it smells like, tastes like, feels like, sounds like. Each sense you include in a description lures readers into your article. A good rule of thumb cited by fiction writers is that each scene should include three senses.
- **Write for the glands.** Use emotion-evoking words, phrases, and descriptions to tickle the adrenal and other appropriate glands. An example of such emotion-evoking wording is the lede for the article "Where the Exotic Meets the Academic" in the online resources for this chapter.
- **Use dynamic quotes.** Chapter 10 on writing news releases emphasizes the importance of using "real" quotes that sound like words a person would actually say. Feature articles should also use real quotes, and they also should be vivid and dynamic and should help move the story along. So, in portraying what people said, pick out those utterances that are concise and meaningful. Pare longer quotes by editing them and using ellipses to capture the dramatic essence of the quote. For example, the physicist Dr. Quark might have said: "Although the results came in over a period of months, after we analyzed them we knew we had a stunning discovery. When we saw the numbers from the initial Fourier analyses come up on the computer screen, we knew we had hit a physics mother lode." For your article, however, you can pare this quote down to the pithier: "We knew we had a stunning discovery," Quark said, and when the analysis was done, "we knew we had hit a physics mother lode."
- **Portray conflict.** Researchers can be reluctant to highlight conflict in their articles, carefully choosing words to minimize any appearance of personal disagreement. However, personal and scientific conflict brings an article alive and humanizes the research. So, tell your readers that Dr. Quark angrily disagrees with Dr. Boson about a theory. Or better, *show* the disagreement: "Quark harrumphed and rolled his eyes at the mention of Boson's name, giving a dismissive wave."

- **Use suspense.** If your article covers research that solved a mystery, use that in your article. Write the article as a timeline, in which the mystery is raised, explored, and solved. Even if there is no mystery, you can structure your article so that it raises a question at the beginning that you answer at the end. For example, introduce a character that encounters a problem at the beginning, and reveal the solution to that problem at the end.
- **Describe action.** Again, show rather than tell. Use your descriptive "camera" to focus on a particular action or event crucial to the story—rather than simply telling the reader that it took place. If you were not present, ask those who were there to give you enough detail—especially including sensory detail—to reconstruct the action or event. See, for example, the lede to the article "Unearthing Ostensible Ancestors," cited in the online resources for this chapter, in which a scene of discovery was reconstructed through an interview with the girl who made the discovery.

Create Hand-Holding, Entertaining Explanations

Explanations of scientific ideas are the heart of your article, so they need to be both interesting and accessible. For complex concepts, think of yourself as a guide leading readers figuratively by the hand along a winding, sometimes rocky path. As would any helpful guide, you would not let go of their hand and would take them by small, manageable steps along that explanatory path. Do not ask them to take large leaps by giving them a lot of information at once.

Sometimes being a good guide means skillful summarizing and digesting: skipping intricate details, caveats, or technical terms—even though they may be near and dear to your researcher's heart. Remember that your purpose is to lure readers along the pathway of your explanation, so that they learn the general concepts of your topic. You do not want their progress blocked by a boulder of technical jargon or the torturous path of an overcomplicated explanation. So use technical terms sparingly and only as necessary, and explain each term concisely as it appears.

Entertaining readers with vivid, even light-hearted, analogies and names are a terrific way to keep them reading. As examples, recall the "artificial dog," "cosmic blowtorch," "anaconda receptor," and "shotgun synapse"—nicknames for scientific objects discussed in chapter 10 on writing news releases. More extended analogies can help readers navigate an explanation of a complex concept. For example, here is an extended analogy that explains biological signaling pathways and how scientists trace them:

> The machinery of every living cell consists of a host of such molecular signaling pathways, like the systems that make up a car's machinery—

> the fuel system, cooling system, electrical system, drive train, and all-important entertainment system that keeps the kids quiet in the back seat. By assiduously breaking one component or another—like, say, taking a hammer to a carburetor—researchers can deduce which pathway each component belongs to.

Use Enabling, Narrative Citations

You cannot use your cherished footnotes in a popular article. To cite another's work, integrate that information smoothly into your narrative text. Include the researcher's name, affiliation, and a sentence or so about the work and how it fits into the topic. If a bibliography is allowed, you can provide the citation and URLs of Web sites with background information, which the magazine can use in the Web version of the article.

Make citations "enabling," says journalism teacher Carol Rogers. They need to contain enough distinctive information so that the reader can find more details. "For example, if information is attributed to scientists at a meeting in San Francisco or scientists at Duke University, those are dead ends," she says. "It could be any meeting in San Francisco, and Duke is a big place. Writing 'Jane Smith, a biologist at Duke University' is better, although the reader still has to be a pretty good sleuth to find Jane Smith and her research."

"Murder Your Children"

Perhaps the most excruciating practice in writing is "murdering your children"—ruthlessly trimming or excising cherished passages of prose that just do not work. Your journalistic children might include the metaphor or simile that does not really aid your explanation, or the anecdote that does not really illuminate the point you are trying to make. Also, be willing to kill off beloved technical jargon or arcane explanations that satisfy you but would leave a lay reader cold. The best way to steel yourself for such literary mayhem is to take time away from a piece of writing and then return for the editing session. In fact, after such a vacation, you might yourself be rather appalled and embarrassed that you once thought that metaphor was apt or that anecdote clever.

Enjoy Yourself So Your Readers Will, Too

Let your writing portray your delight in your topic, says Sally Maran: "Scientists should let their passion come through. They should let the reader know why they are so thrilled with what they do. And humor always helps. Maybe it's an outrageous pun or a funny story, but don't back off from humor. If you can find something amusing in what you do, that is great."

Develop Magazine Illustrations to Tell Your Story

Images are critical to most popular articles, and you will endear yourself to the editor and the art director by providing ideas for visuals. They need not be polished, but only "art scraps"—photos, sketches, or other materials that the art director can build on. Even your own hand-drawn sketches of ideas for artwork can be invaluable, no matter how primitive. Slides from your talks can be useful, especially if you have heeded the advice from chapter 4 on developing good visuals. Also, suggest photo possibilities that you uncovered in your research and interviews. The art director will use such materials as a basis to assign photography or commission illustrations.

Besides finding use in the article, such images can also influence the article itself, points out Reid: "We often have these conversations around the pictures that lead to the text being revised," she says. "A picture is a concrete representation of a concept, so it can enable us to go back to the manuscript and ask good questions about whether the words explain that concept adequately." Do not limit yourself to still images, but also think of animations, video, and audio that can accompany your article. Magazines often use such visuals for the Web versions of their articles. For example, take a look at the Web sites for *American Scientist* and *Scientific American* to see online visuals they use with their articles. You can also post such visuals on your own Web site and use them in your talks, with permission from the magazine.

17

Author Popular Books

"Gee, you should write a book." You might have heard such an offhand comment from a friend or harbored the ego-flattering dream of being an acclaimed author that has led you to consider writing a book about your field. But before you leap into this huge commitment of time and energy, you should understand the pros and cons—especially the cons. And, understand the frustrations of the publishing process.

The Pros

A popular book will increase your visibility as an authority in your field, making it easier for you to get the attention of important lay audiences and even your colleagues. A popular book will bring invitations to write articles and lecture that otherwise might not have come. "Before I had written books I don't remember ever getting invited to give a lecture, other than based on my technical work," says Henry Petroski, author of highly popular engineering books, including *The Evolution of Useful Things* and *The Pencil.* "Now I probably get one invitation a week for a major lecture."

A popular book could also make you a de facto public spokesperson for your field, your opinions sought after by journalists and legislators. A popular book also enables you to explain your field in a way you believe it should be explained. A book is, indeed, a bully pulpit, enabling you to define the mysteries, issues, and controversies in a way that no article or collection of articles can. And of course, a popular book is a great way to impress your family, and it gives them something to boast about at family gatherings when cousin Fred the lawyer is blathering on about his big cases.

The Cons

A popular book will not likely make you rich, or even sort of rich. The vast majority of such books barely earn back their advance, which typically runs in the four figures for a nonfiction book. In fact, the per-hour pay for writing a book—given the huge amount of time required—typically ranks below even the pay for collecting garbage.

Also, the time spent writing and publishing a book will take away from your research and professional writing. This time commitment continues after the book is published. To make the book well known, you will have to spend considerable time promoting it. So, do not consider writing such a book unless your research career is well established, or you are a compulsive writer.

Writing a book requires the discipline to spend years researching, writing, and going through the laborious, frustrating publishing process. Also, says author and *Science News* editor Tom Siegfried, "You have to have passion and the motivation. If you don't have a real passion to tell your message to people because you think it's worth people knowing, then it's hard to sustain the quality of writing and intensity of working you need to stick with such a big project."

The visibility from a popular book can make you a sitting duck for critics. They include those who disagree with your positions on research and policy issues. And they include editorial snipers who simply enjoy taking potshots at public figures. So you will have to prepare yourself for criticism, both by taking great care to make your writing accurate and by developing a thick skin.

Writing a popular book will thrust you into the capricious, frustrating world of publishing. Unless you are lucky enough to find a supportive publishing home, your work will likely be arbitrarily rejected, and at some point sloppily edited and published. Every author has war stories to tell about foolish editors. Henry Petroski's favorite silly-editor story is about the proposal for his best-selling *The Pencil*. "One editor gave my proposal to his assistant, and the assistant circled on some sample pages the word 'pencil,' to point out that the word occurred a lot. The editor explained to me that was a basis for their rejection. Well of course, the book was *about* the pencil, and there aren't too many synonyms for it!"

Even an excellent nonfiction book is at the mercy of fickle trends. Your book proposal might be rejected as being on a topic that is out of fashion at the time. Or, the topic of your finished book might go out of fashion before it is published. The arbitrariness of publishing decisions is well illustrated by the serendipitous path to publication of Robert Cooke's book *Dr. Folkman's War: Angiogenesis and the Struggle to Defeat Cancer*, although this story has a very happy ending. He proposed a book that would cover the life and work of the late Harvard researcher Judah Folkman, whose discoveries about the role of angiogenesis—the growth of blood vessels—in sustaining cancers had achieved renown after long being controversial. Cooke's reporting on Folkman had led the scientist to agree to cooperate with him on the book. The book proposal had been rejected by one editor, and Cooke's agent was preparing for a long, difficult slog through publishing houses. Serendipitously, just then the *New York Times* ran a front-page story touting Folkman's work. The article quoted Nobelist James Watson as declaring that "Judah is going to cure cancer in two years." The article and the quote set off a storm of media coverage. The result: a furious bidding war among publishers for Cooke's book, resulting in a handsome advance. But such stories are rare. More common is the long slog through the publishing process that involves developing a proposal, pitching an agent or editor, responding to committee reviews of the proposal, negotiating the contract, editing, proofing, and marketing.

Do You Have Something New, and Marketable, to Say?

You need to have something new and significant to say before you should consider writing a book, emphasizes journalist/author Keay Davidson. "That will distress some people because they think 'I have been a physicist for twenty years, and I can write a book about my experiences,'" he says. "But the book market is ruthless and difficult, with most books not selling and not even getting reviewed. The ones that do attract attention are the books that have some kind of clear thesis." Davidson's books are good examples of clear, engaging theses. His *Carl Sagan, a Life* offers insight into one of the most famous, and sometimes controversial, scientists of his time. And his forthcoming book *The Death of Truth: Thomas S. Kuhn and the Evolution of Ideas* explores the life and work of the most influential historian of science of all time.

Even if you have something new to say, will your ideas be marketable? So, think about the market before you even begin to write your book. Identify the potential audiences for your book and how you will reach them. Ask yourself what will compel them to buy the book. If the audience is small, or the motivation for buying the book is weak, perhaps you do not have the makings of a popular book. Perhaps you have a professional book or, if the topic is too narrow, an article for a specialty magazine.

A marketing-first approach also can help shape your book. It might suggest a different approach or additional material that will make the book more useful and/or interesting to readers. For example, *Explaining Research* began far more modestly, as only a booklet for researchers on dealing with media. However, as the project evolved, it became apparent that researchers needed a far more comprehensive guide on explaining their work to all audiences.

Motivating people to buy your book is critical. You might believe that readers will flock to your book just because you are a prominent researcher. But readers really do not care much about your credentials; they care about what you have to say. You might also believe that people should read your book because it is good for them. This "eat-your-spinach" approach to popular books, and for that matter all lay-level communication, does not work. As with your other communications, you will attract an audience not just because your topic is important, but also because you make it interesting.

Write the Book Yourself or Collaborate with a Writer?

Being a compulsive writer is one reason for writing a book, and it is a legitimate one. Some people, me included, *must* write. You are born to write if, as Gloria Steinem once said, "Writing is the only thing that, when I do it, I don't feel I should be doing something else." Henry Petroski is a born writer, recalling that "from graduate school onward I was always in two camps. I was writing poetry really late at night, after the day was done. During the day I would write technical articles and reports, proposals." Petroski says he found that his writing evolved naturally from poetry to essays to books.

Even if you are a born writer, you will still "suffer" for your art. Pulitzer Prize–winner Jon Franklin warns about scientists who aspire to write. "They have this romantic notion of what being a writer is, but it is every bit as reductionist as science. Writing is not fun; having written is wonderful. Having written is something I am totally addicted to."

However, if writing is not an addiction for you, and only a necessary onerous chore, consider collaborating with a professional writer on your book. "I would say to ninety-nine percent of researchers who want to write a book, you probably are not going to be any better at writing a book than I would be at doing molecular biology," says journalist/author Paul Raeburn. "So hire a writer. Let the writer write the book, while the scientist provides the information through interviews or journal articles. The role of the scientist is to make sure that the content reflects what the scientist wants to say, but the writer is the expert at putting that into language that will speak to a broader public than scientists normally directly address."

Such coauthors are not merely copy editors, emphasizes Davidson. "The coauthor is not just there to dot the i's," he says. "A true coauthor is someone who wants to help them tell a story.... They have to be willing to listen to the coauthor when they tell them 'This is just incomprehensible; nobody will have a clue what you are talking about.'" The harsh reality is that unless you seek to collaborate with a professional science writer who has experience with authoring books, your book will not likely be written. Even with the best of intentions, your inexperience as an author and the demands of your professional duties will doom the project.

If you seek a collaborator, do not limit your assessment to reading what a writer has written, but interview the writer, advises *New York Times* science reporter Sandra Blakeslee, author and coauthor of many books. The interview should determine personal rapport and also basic writing ability, even with prominent writers, says Blakeslee. "Don't just read their published pieces, because some very prominent writers are terrible," she warns. "They have gotten heavy editing from their news organizations."

Also, she says, do not expect to collaborate long-distance, but schedule personal work sessions every few weeks, lasting days. And even though you are working with the most industrious writer, do not expect to simply feed the writer ideas and relax, says Blakeslee. "If you are the scientist, you can not assume that you are going to have this science writer who is going to waltz in, read your mind, put it down in your language, get it right and the way you like it, and you are not going to have to do any work," she says. "You are going to have to work your butt off. And it is going to be as hard as if you wrote the book yourself, but it will be much, much, much better."

For one thing, she notes, researchers likely do not understand such writing issues as pacing, style, and the use of anecdotes and colorful examples to energize a narrative. More basically, she says, researchers often do not understand how a book's theme and organization can evolve as the book progresses. "I could guarantee any structure you started out with is not what you will end up with. It changes every single time, and chapters will flip, segues will flip. It is just because as the story unfolds at book length, it will change." Thus, says Blakeslee, researchers must be flexible about changes in the book as it evolves.

You will also have to give up your cherished academic writing style, says Blakeslee. She recalls her collaboration with neurologist Vilayanur S. Ramachandran on *Phantoms in the Brain*. "Rama had a very stiff Oxbridge-type training, so I would send him a chapter, and then he would rewrite it and take all of my nice declarative sentences and combine them and add verbiage. I would get it back, and I would chop them up, and I would send it back; and then he would send it back; and it was like a tennis ball we just batted back and forth." She would tell Ramachandran "It is complex material; you have to have short declarative sentences; go easy on your readers; don't make them struggle."

Different scientists work with writers in different ways, notes Blakeslee. She recalls that working with one collaborator required many months of talking and many drafts for the book to emerge; but in another case a book emerged quite readily from recordings of her coauthor's undergraduate lectures. Despite the complexities and frustrations of collaborations, researchers willing to cooperate with the writer can arrive at a text that portrays the researcher's "voice" and explains his or her ideas accurately and compellingly.

Another type of "collaboration," is one in which the writer is the sole author on a book about your research. Such was the case in Cooke's book on Folkman. For his book, Cooke conducted some 18 extensive interviews with Folkman, as well as with research colleagues and people who knew about Folkman's early life. In such situations, the writer is acting as a journalist, with more autonomy. So, you should be willing to respect that autonomy. Cooke describes the interaction: "I would come in with questions that I needed answered for the next chapter, and he would go through them and answer them. One of our rules was that I would let him look at the book only for the sake of accuracy. It was my judgment as to what would or would not be in the book."

Train for a Book

Whether you write the book yourself or collaborate with a writer, you can undertake to train, literally, for the marathon process of authorship. Here are elements of that training regimen:

- **Get experience explaining your work.** "Practice at every opportunity explaining what you do to nonspecialists," says Raeburn. "Give talks at the library or any other public venue. Do whatever you can do to get a sense of how to try to convey your ideas to somebody who doesn't have a technical background."
- **Read good writers.** Read popular books in your field, as well as other fields. The online resource section at *ExplainingResearch.com* lists some of these books. Do not just read for content, says Petroski. "I learned to be sensitive to words through reading first for structure and effect. I get a lot of pleasure out of having read a really elegant passage or sentence,

whether it was the sound or the way the words fit together," he says. "Even before beginning to read literature or science, I remember reading the *New York Times* sports section, not just to get the scores, but to read really good sports writers."

- **Start short.** Start by writing short pieces before you graduate to long articles. Write long articles before you graduate to a book. Petroski took this graduated path, and he says that "all those years of 'apprenticeship' taught me that you have to be your own severest critic." Thus, he can better judge his own writing, even to the extraordinary extent of abandoning entire book projects when they are not working.
- **Become a literary squirrel.** Books require an enormous amount of research, and you will make this process easier if you get into the habit of squirreling away ideas and material as they cross your desk or your computer monitor. However, do not simply toss material into an unorganized file and risk losing it. Consider using a database, such as the popular askSam, which can store and organize Web pages, e-mail messages, pdf files, and text documents. Or, consider Nota Bene, a powerful integrated word processor/database manager.
- **Cultivate "graphophilia."** Develop a writing habit by setting aside an inviolable time to write. For example, Petroski got into the habit of writing late at night, "when I would go to my desk at home and either write a poem or write an op-ed piece. My wife likes to go to bed early, and the kids were small so they went to bed early. I had the house to myself." He did this even though "I was very tired. There were many nights when I would literally fall asleep at the desk."
- **Read books on writing.** As mentioned previously, Franklin's *Writing for Story* and the writing guides published by Writers Digest books are very useful. See the online resources for this chapter at *ExplainingResearch.com* for a list of useful books on writing and marketing.

Find Agents and Publishers

Agents and publishers can be difficult to attract, given that they are inundated with material from would-be authors. Here are tips to help:

- **Explore *Literary Marketplace* and *Writer's Market* to find publishing houses, editors, and agents who handle the kind of book you are planning.**
- **Identify the publishers of other books on your subject and contact those publishers to determine who edited the books.** They are key targets for your book proposal.

- **Contact authors of books that you admire and find out who their agents are.** If you know an author well enough, he or she might be willing to recommend you to their editor or agent. Such recommendations can be crucial, because editors and agents often ignore would-be authors who do not come recommended.
- **Get to know agents and editors in social situations.** Editors often scout scientific meetings for potential authors. You can learn a lot about the publishing business in discussions over a lunch or dinner. You can also refine your book idea by pitching it to them.

Prepare a Query Letter and Book Proposal

Your first contact with an agent or editor will usually be a query letter. This should be a compelling, concise document, consisting of a one- or two-page description of your proposed book, why it is unique, why you are qualified to write it, and why it will attract a wide readership. The potential readership is crucial. You must convince an agent or publisher that your book will find a significant audience and turn a profit.

Once the agent or editor has responded positively to this literary appetizer, follow up with the entrée: a substantive book proposal. The most persuasive proposals typically comprise

- A one-page synopsis of the book that explains the subject, your approach, special features such as illustrations, and how it will benefit the reader. An attention-getting title is helpful, although you will not have final say on the title.
- A market analysis that includes the size and demographics of target audiences and how you will publicize and promote the book to them.
- An analysis of competitive and related books and how yours will be different.
- A proposed table of contents.
- A chapter summary, in which each chapter is briefly summarized.
- Sample chapters, usually three, to illustrate your writing ability and tone.
- Popular articles you have written, especially those that cover the book's topic.
- Popular articles about your work and review articles that cite your research.
- Detailed information on your professional and writing qualifications, with a link to your Web site and your e-mail address.
- A self-addressed, postage-paid envelope for return of your material.

Edit and proofread the proposal obsessively. Make sure it is compellingly written and with absolutely no technical verbiage. Ask others to read and critique it, perhaps your public information officer (PIO).

One particularly helpful book is *How to Write a Book Proposal* by Michael Larsen. The online reference section lists other such books and links to information on book proposals.

Become a Publicity Hound

Assuming you have run the sometimes painful gauntlet of selling your book to a publisher, then writing it, and then getting it published, your job is by no means finished. The next step is to shamelessly promote your book, a process that you may be acutely uncomfortable with, since you are a refined, respectable researcher. Nevertheless, unless you effectively publicize your book, the most brilliantly written tome will go unread and unappreciated.

The cold truth is that your publisher will not publicize your book, at least not in any significant way. A typical publisher throws a book into the maelstrom of the marketplace—issuing a perfunctory news release, sending it to some reviewers, and giving it a few months to sell or flop before relegating it to the backlist. Even so, you should still supply your publisher with all the information needed to market your book, to the extent that they are willing. Your publisher will likely supply you with an author questionnaire eliciting reviewing, marketing, and publicity possibilities. Fill it out as thoroughly as possible.

Given publishers' relative lack of marketing, expect to spend just as much time marketing your book as you did writing it and getting it published. If you do not wish to take time for such marketing, expect the book to become little more than a useful addition to your CV.

As discussed previously, you should have begun thinking about marketing your book before you wrote it—identifying your audiences and how to reach them. And, presumably you wrote the book to target those audiences, giving them information they are willing to spend money on.

Here are tips on marketing your book once it is published:

- **Read good books on marketing.** They include *1001 Ways to Market Your Books* by John Kremer and *Guerilla Marketing for Writers* by Jay Conrad Levinson, Rick Frishman, and Michael Larsen. Others are listed in the online resources for this chapter.
- **Shamelessly solicit blurbs.** Contact the top people in your field, asking them to write a brief laudatory blurb on your book. If they have written books, they will appreciate the importance of giving you a good blurb.

In fact, you can even suggest blurbs. Particularly good prospects for blurbs are people cited in your book.

- **Take full advantage of Amazon's marketing tools.** It surprises, even stuns me, how many excellent science books did not have the whole book posted for Amazon's "Search Inside the Book" feature and did not launch an Amazon Connect blog or use Amazon's other marketing services. To help you take advantage of the power of Amazon marketing, read such books as *Aiming at Amazon* by Aaron Shepard or *Sell Your Book on Amazon* by Brent Sampson.
- **Use social networking.** Launch and maintain a blog, e-mail newsletter, and a Web site with the URL *YourBookName.com*, or as close as you can get.
- **Do media interviews.** Work with your publisher to identify radio and TV shows on which you can appear. If your publisher is not cooperative, make contacts yourself or hire a publicity firm specializing in book publicity. Your PIO might also offer advice or help with media contacts. The techniques for giving radio and television interviews covered later in this book will help you be effective.
- **Take advantage of freebies.** Finally, and perhaps most heartening, is that most things you can do to market your book are free or inexpensive. For example, marketing on Amazon is free, as are most of the marketing techniques outlined in the books listed above.

18

Become a Public Educator

Even though your prime duty is to your research, fully serving that research means serving your field as a whole. And an excellent way to serve your field is to become a public educator. What's more, besides benefiting your field and the people you teach, public education also brings professional benefits. You will become a better communicator, able to advocate your work to important constituencies—such as the vice president who is deciding whether to fund a new building to house your laboratory. You also will become more visible, which will not hurt your career, either.

So, here are ways you can become a public educator—by giving talks, teaching K-12 students, mentoring young people, and working with science museums.

Give Public Talks

You might be surprised at the many venues to give public lectures on your field. Your institution, your scientific society, Sigma Xi, and AAAS all offer golden opportunities to reach a broad audience:

- **Give public lectures at your institution.** For example, Harvard sponsors the Longwood seminars, which are "Mini-Med School" classes for the general public. You need not be a senior researcher to participate in your institution's public lectures. Harvard's public Science in the News lectures are given by graduate students and cover hot scientific topics and their ethical or social ramifications. Those lectures can reach far beyond your institution, for example, being posted as audio or video on Apple's iTunesU service or on your institution's Web site.

- **Register with your institution's speaker's bureau.** Such bureaus advertise your availability to civic clubs, school assemblies, and other venues. Also, if you are at a university, your alumni office would likely welcome your willingness to talk to alumni groups. You might combine a trip to a scientific meeting with an appearance at a local alumni club.
- **Give a science museum talk.** Such a talk can be a particularly good way to get started giving public lectures, says Tinsley Davis, executive director of the National Association of Science Writers who organized talks for the Boston Museum of Science: "In talks at the museum, for example, everybody in the audience loves science, so it is different than most audiences," she says. "The audience is small, and scientists can bring their friends and families. It is a soft place to land for a first talk."
- **Participate in the Science Cafés program.** These small, informal meetings give you a chance to meet with people in pubs and coffeehouses to talk about current science topics. This is a painless way to get started talking to the public, because Science Cafés typically require little preparation and give you immediate, intimate feedback from interested people. NOVA ScienceNow and Sigma Xi provide information and coordination at *ScienceCafes.org.*
- **Participate in AAAS public programs.** Talks at the AAAS meeting, or as part of the AAAS Center for Public Engagement with Science and Technology, offer a chance to not only convey your research to a broader science-oriented audience, but also to educate the public through the media. AAAS public programs director Ginger Pinholster calls the annual AAAS meeting "the Olympics of science conferences." She points out that the annual meeting attracts about a thousand press registrants from around the world. You can either give a talk at a symposium or propose one, either on your own or with colleagues in your field. Contact the annual meeting at *meetings@aaas.org* to explore how to participate.
- **Advise media on science and technology.** Media, ranging from your local journalists to Hollywood filmmakers, often welcome advice on portraying science and technology in their articles, television series, and movies. For example, the National Academies has formed the Science and Entertainment Exchange to link the entertainment industry with scientists and engineers.

Whichever venue you choose, expect to audition for the people who run such programs by chatting with them about your subject in an accessible, compelling way. For example, Davis outlines what she looked for in a public speaker when she recruited speakers for the Boston Museum of Science. "We wanted someone who is passionate about their work, knows their subject, and is a good communicator," she says. "We were looking for scientists who could make grass-growing interesting." Davis says your audition will be more successful if you offer ideas for good videos and other visuals to illustrate your talk. And once you have developed your presentation, expect the organizer to want to review it, to make sure slides are understandable and present scientific concepts clearly.

Of course, in developing your presentation, use the same presentation skills outlined in chapters 3 and 4 on giving compelling talks and developing effective visuals.

Work with Local Schools

Explaining research to an audience of squirming kindergartners presents the biggest challenge to any researcher-educator. But the effort is worth it. Not only will it give you a baptism of fire in teaching you how to reach a lay audience, but how to do so with some pizzazz.

Beyond teaching tykes, there are many ways you can involve yourself in your local schools. The NIH report "Scientists in Science Education" offers an excellent discussion of the possibilities, including

- **Visiting a science classroom**
- **Providing teachers professional development or training**
- **Helping with informal science education programs, such as school programs at science centers**
- **Working with parents and school boards**
- **Developing instructional materials**

You can contact a local school principal or teacher directly to offer such services, and/or work through a partnership run by your own university or laboratory. A prime example of such partnerships is the school programs offered by the Salk Institute—including a summer enrichment program, a high school science day, and a mobile science laboratory.

If your institution or professional association does not have such K-12 education programs, encourage it to develop them. Not only do such programs inspire children's interest in science and enhance community relations; they also train researchers in lay-level communications. "I have had many scientists tell me the process of working with high school kids, or even younger kids, forced them to think hard about how young people learn," says Duke communication director David Jarmul. "And as a result, they came back to their own undergraduates and graduates and postdocs as better teachers." Such skills also extend to other audiences besides students, says science communicator Lynne Friedmann. "Those foundations and non-profits with well-developed programs that reach out to children find that their scientists do a much better job of getting information out to other audiences as well. When they participate in career days or talk to kids, they learn communications skills they might not learn in talking to other audiences."

A good way to learn about science teachers and the issues they face is to explore the Web site of the National Science Teachers Association. More locally, involve yourself in an NSTA state chapter. Also, NIH offers aid in developing teaching material. For example, NIH's Science Education Partnership Awards funds grants for innovative educational programs.

If you decide to become even more deeply involved in science education, you might even develop a program yourself, with funding from the NSF Directorate for Education and Human Resources or from such private foundations as the Howard Hughes Medical Institute.

Mentor Young People

Mentoring a science student can give you a profound impact on the future of a promising young researcher. You could contact your local school guidance counselor to recommend promising young people who might want to intern in your laboratory. And the high school might even be willing to give the student class credit for such participation.

You might also volunteer at a local career day to talk about being a researcher. Offering yourself as a role model, especially if you are a woman or underrepresented minority, can have a profound effect. Friedmann cites the effects on girls that was documented in one study of career events featuring women scientists. "When girls were asked to draw pictures of scientists, they would invariably depict old white men in lab coats," she says. "But after a career event featuring women scientists, the girls would begin drawing female scientists and not the white males."

Even giving an informal talk at a lunch with young people can have a considerable impact, says Tinsley Davis. She cites as an example, the museum's Women in Science day. "Women scientists would give short talks and have lunch with these preteen girls, and just the sight of a researcher in tennis shoes and a t-shirt would act as a powerful role model," she says.

A good source of mentoring information for women scientists is the Association for Women in Science book *A Hand Up: Women Mentoring Women in Science.*

Work with Science Centers

Science centers such as museums and zoos are premier gateways for reaching the public with information on your field. According to the National Science Board report *Science and Engineering Indicators 2008*, about three in five Americans visit science centers each year.

In working with a science center, you might give a talk or demonstration, participate in a panel discussion or job fair, or serve as a volunteer explainer. Or if you are willing to involve yourself more deeply, you could help develop an exhibit. All these activities can have an important impact on public science education, and as outlined below, offer many professional benefits.

The Association of Science-Technology Centers Web site offers a good overview of science centers and can help you identify those you can work with. Another good resource is the Center for Informal Learning and Schools. In particular, the center offers a comprehensive resource Web page.

The Many Attractions of Science Centers

Maureen McConnell, a veteran Boston Museum of Science exhibit developer, cites numerous reasons for working with science centers:

- **Help recruit the next generation of scientists.** "We try to make the work that scientists do seem exciting and relevant and worth doing, so that kids can see themselves in that role," she says. And if you are a woman or underrepresented minority, your participation offers an invaluable role model.
- **Connect science with the rest of society.** "Scientists all too often see their work as separate from culture, whereas it is really a cultural pursuit," says McConnell. Particularly valuable are scientists who "understand that the work they are doing is important to the future of the planet.... They realize that what they're doing is of great significance and they want to increase their outreach."
- **Potentially have extraordinarily wide impact.** McConnell points out that many exhibits last for as long as a decade, as traveling exhibits that visit many science museums.

Besides these benefits to society, working with a science center also benefits you personally and professionally. Developing an exhibit can do the following:

- **Teach you valuable communication techniques.** These might include techniques for creating multimedia, computer simulations, and hands-on experiences to convey scientific concepts. Such interactive exhibits, says McConnell "take people through an experience that engages their attention more deeply and for a longer period of time, than is normally the case for an article."
- **Give you a better sense of two-way communications with audiences.** Developing an exhibit will teach you a more accessible style to quickly capture and engage lay audiences than the one you use with your colleagues and students. "Even those researchers who teach just don't think about the difference between having students whose parents are paying for them to learn—and whom they have for a semester—versus someone who is walking by whom you are trying to snag for three or four minutes," says McConnell.
- **Provide useful teaching materials.** You might adapt exhibit content as teaching tools for your classes, your Web site, or even a textbook. Museums often give away computer software or other materials created for an exhibit—especially if the request comes from a researcher who helped develop it.
- **Enable you to synergize your research communications.** You can often adapt your research videos, animations, or other content for an exhibit, and vice versa.
- **Give you public exposure.** The launch of a new exhibit brings publicity and other communications that present you as a spokesperson and advocate for your field.
- **Meet your grant's outreach requirement.** For example NSF's Broader Impacts Review Criterion for grant proposals asks that applicants include an educational component in their grant proposals. One way to meet that requirement is to develop a collaboration between your research project and a science center on an exhibit. You might even cite the center as a collaborator on your grant proposal, increasing your funding chances. Conversely, the science center can use your support as a springboard to seek other funding sources and partners for exhibits.
- **Highlight your institution and your department.** Just as working with local schools enhances your institution's reputation as being civic-minded, so does helping your local science center.

- **Add new nodes on your professional network.** Working with a science center can be the gift to yourself that keeps on giving. You might meet new research collaborators and establish a connection with the center's professional staff, who themselves have a broad range of contacts. "I am not done with my relationships with a scientist when the exhibit is done, because my broad interest is trying to get the word out about science," says McConnell. "So anytime I can see a possible connection among scientists I know, I will try to make it." For example, when McConnell collaborated with ecologists on an exhibit about forest canopies, she discovered that the researchers were seeking a place to teach their colleagues to work in the forest canopy. The group could not find an appropriate facility until McConnell connected them with the Worcester, Massachusetts, EcoTarium, which already had a canopy walkway.

Are You Science Center Material?

Working with a science center does take a particular personality type—one that can tap into the enthusiasm for science that first attracted you to research. McConnell has her own test: "I will ask a scientist 'How did you get into this field? What did you love most when you were a kid?'" she says.

> In developing an exhibit, I'm trying to create the same kinds of hooks that snagged the scientist as a kid. If they can't remember their childhood, I don't work with them. I find that those people just can't adjust to what it is you are trying to do when you make an exhibit. If they can't remember it, they can't create it.
>
> I expect them to be eccentric and kooky and totally nuts about their thing and maybe not be able to talk about anything else.... I love them for being that passionate, even if it is somebody who talks to you in a monotone and can't dance, they are passionate.

Says Carol Lynn Alpert of the Boston Museum of Science, audiences should "come away inspired by the energy and the quest of science rather than remembering all the facts that you put up on the slides for them.... When we look for a scientist to participate in face-to-face encounters, we are really looking for someone who puts themselves into the story; someone who relates their story to personal experience; someone who can share with us what it is really like on the inside doing science."

You emphatically do not have to be a scientific "elder" to work with a science center, points out Alpert. "The young people in our audiences can sometimes relate more easily to graduate students and young people in research than they do, say, to the well-known Nobel laureate."

How Much Time?

Developing an exhibit can take years—including a year or so to develop a grant proposal and secure funding and another year to develop the exhibit. So, expect to commit yourself to a project for that long. However, that commitment will not be onerous, emphasizes McConnell. "Exhibit developers are sensitive to the time constraints of scientists," she says, "They recognize that scientists are quite busy. So, I tell scientists that I will not take any more of their time than they feel they can give."

For example, your task might be only to write letters to aid fundraising for a project, or review proposals and exhibit plans, says McConnell. Or, you might attend periodic advisory committee meetings, which can be coordinated with travel schedules for other purposes. So, when forming a relationship with a science center, feel free to define your time constraints, and the staff will honor those constraints.

How to Get Started

To get involved with a science center, you do not have to match your interests with existing exhibits, advises McConnell. "Pick a museum, approach them and tell them what you're doing," she says. "Tell them 'I think I have some great stuff here that people will be interested in. What are some venues for telling this story?' Then you can work with them to decide the best way to get involved—whether exhibit development, talks, or service on an advisory committee. Be enthusiastic, and the center staff will usually bend over backwards to help you."

19

Persuade Administrators, Donors, and Legislators

Because they hold the purse strings, administrators, legislators, and donors rank among the most important audiences for explaining your research. They share many characteristics as audiences. For example, they all

- are usually not scientists or even versed in the issues that concern you,
- have limited time and energy to give to your communications,
- are the object of much lobbying other than yours and thus are often wary, even suspicious, of motives,
- prefer information as concise "nuggets"—compelling summaries and examples that emphasize how your work affects them and/or their constituencies, and
- must justify their decisions to those constituencies, whether governing boards, oversight committees, or voters.

You may already have the content to help you reach these audiences, in your news releases, feature articles, PowerPoint presentation, Web site, and other materials. This chapter outlines how to use that content effectively.

Engage Administrators

To reach your institution's administrators, first put yourself in their place. What institutional issues are they addressing at the moment? Are they trying to launch new research programs? Where are they in the budget cycle? What funding and/or facilities problems are they facing? What personnel problems?

Take into account that you live in different worlds. You might worry about fixing that DNA analyzer or getting quality data from that particle accelerator. Administrators are just as concerned with that looming foundation proposal or the outrageous construction bid on that new building. Get into their mindset and you will have an easier time connecting with them.

Also, what are their most important constituencies, and what do they need to communicate to them? For example, university administrators typically must produce reports to a board of trustees, as well as trustee committees overseeing different aspects of the university. If you read such reports, you can figure out how to adapt your research explanations to be useful in the reports. Consider how information about your research will resonate with your administrator's audiences. For example, from her tenure as a communicator at the Idaho National Laboratory, Duke's Deborah Hill recalls how research information needed to support objectives of the federal agencies:

> It is generally accepted that university researchers can pursue projects just for the sheer intellectual delight of the pursuit. But when people at national labs do media outreach, they need to take special care to show the connection to something significant to government—whether that happens to be the whim of the day or a national issue. Program managers will take those stories up to the Office of Management and Budget or the White House. And those research projects may not always be the sexiest in the sense of producing a breakthrough; they may be ones that can show such a connection.

So, in communicating with administrators at national laboratories, "You can't just focus on the very narrow results of any particular paper; you always need to present your work in a bigger context," says Hill.

Similarly, corporate scientists need to emphasize how their work contributes to the company's business, says Seema Kumar, vice president for global communications at Johnson & Johnson. "Even though some of the commercial folks don't have a scientific background, they know enough about the field that they don't need a basic explanation," she says. "What they are looking for is: how well does this drug work? Is it safe? Is it effective? What is the market?"

Once you understand the needs and interests of your administrators, plan your communication strategy accordingly. Some tips:

- **Meet with them periodically.** But only do so if there is a specific issue that needs to be decided.
- **Be social.** If the political/social climate is right, schedule lunches and other informal meetings just to keep up with the issues and developments of the moment. Not only is good science often hatched in the lunchroom, but also many good management ideas.
- **Keep them updated.** Without pestering, keep them constantly informed of your research progress. Send articles, media clippings, and other materials that they might find interesting or useful for the communications they must do.
- **Copy them on congratulations.** When you send congratulatory notes to colleagues about new grants, awards, and so forth, copy your administrators. Such inclusion helps them stay current with the institution's achievements and might prompt them to pass the note up the chain of command. If you are a department chair, copying congratulatory notes gives you a chance to remind the administrator of your department's achievements.
- **Congratulate them.** Administrators are just as proud of such achievements as procuring grants and launching new programs as you are of discovering a new gene or creating a new astrophysical theory. But while you might receive kudos for publishing a seminal paper, administrators are seldom recognized for even major administrative achievements. Once, for example, the provost of a university where I worked approached me absolutely beaming, because he had just procured a major grant after years of work. However, his triumph would only be shared with other administrators, because the brilliance of his coup could never be really explained to the university community. The formal announcement of that very large grant would come only in the dry news release, complete with the perfunctory "We are very pleased to have received this grant" type of quotes. The provost wistfully asked if I could do a feature story on how he got the grant, which I had to decline because it would have come across as self-serving. Sending a congratulatory note to this provost in the hour of his proud success would have been most appreciated.
- **Thank them for their help.** Too many researchers will lobby administrators intensively for funding or facilities, only to fall mute and return to their laboratories once the administrators have given them what they wanted.

Persuade Donors and Foundations

As you probably well appreciate, private donors and foundations play an invaluable funding role—offering support for young researchers, as well as senior researchers who have untested ideas that government agencies would not touch. However, donors and foundations are also more individualistic, sometimes even eccentric, than the NIH or NSF. So, your communications with donors and foundation officers also will have to be more targeted.

Your development officer is your best ally in working with donors. So, get to know him or her and the office's policies and procedures for communicating with donors. For example, development officers usually coordinate approaches to particular donors—ensuring that donors are not bombarded with multiple diverse pitches from the institution and that pitches match donor interests. Also, your development officer knows best which donors are interested in your area, how much they will likely give, and how best to shape communications with them. Even large private foundations, which usually explicitly state their funding priorities and procedures, often have informal interests and agendas that development officers track.

Development officers will need you to supply donor-worthy communications materials—news releases, feature articles, videos, podcasts, Web and blog content, multimedia packages, lay-level PowerPoint presentations, and your concise "elevator speech" summarizing your work, described in chapter 23. However, this content should be tailored to each prospective donor. For example, even if you do excellent research on a range of cancers, for a donor interested only in breast cancer, you would concentrate your information on that cancer. And shape your content to fit the donor's funding history and activities. For example, when I was at Duke, its development officers wished to cultivate for further gifts a prime donor who had given the "naming gift" for a major research building. So, they regularly sent a tailored report on the research discoveries made in "his" building. That report was easily compiled from news releases and features I had already done on researchers in that building.

To make sure donors receive the latest information on your research, keep development officers apprised of the progress of your work. For example, even though they see news releases and in-house articles on your research, they might miss even prominent media articles. Also, the officer may miss the latest content added to your Web site, including videos, podcasts, and so on.

Even though your content is compelling, personal contact is most persuasive to donors. They want to learn not only about your research, but also about you as a person. So, aim your presentations accordingly. Include anecdotes that reveal your passion for your work and for its ultimate goals—whether curing disease, building earthquake-proof bridges, or seeing galaxies at the edge of the universe. Donors also want to know how their support will make a difference. So, be prepared to talk about the specific experiments and equipment a donor's funds will support, as well as the ultimate applications of that work.

Donors may have very personal reasons for their interest in your work, says science communicator Lynne Friedmann. "Many times donors will have a family member that has a condition, or they know someone with it, and you need to know that. Also, people with big bucks sometimes donate to causes because of how it raises them on the social ladder in a community. So, while you should have your elevator speech prepared, also be able to give them a reason for donating that they can articulate to their friends and that might get other people in their circle involved."

In communicating with donors who have given gifts, make sure you know whether they prefer e-mails, phone calls, newsletters, Web URLs, or visits to your laboratory. Also, find out how often they want to be contacted. Your development officer can help you determine such preferences.

Donors are most appreciative of prompt, personalized acknowledgment of their gift and follow-up reports on how that gift has made a difference. As indicated earlier, donors—like administrators and legislators—prefer their communications to be concise. A one-page letter or short e-mail, perhaps with a link to a news release or feature article, is usually the best. And do not contact donors just to ask for money. Drop them a note when you run across a news article or event they might be interested in, even if it does not involve your research.

Lobby Legislators

Researchers should not be surprised when the budget for their funding agency gets cut—whether DARPA, DOE, EPA, NIH, NSF, ONR, or USGS. After all, budgets for these agencies are decided by legislators who are not researchers and who see far too little public advocacy by the scientists affected. Nor should researchers be surprised when so many public officials take less-than-"scientific" positions on issues such as global climate change or planetary exploration. "Scientists are not anywhere near as involved in public policy in this country as they should be," declares Duke's David Jarmul, who has worked at the National Academies.

> In Congress, there are only a handful of physicians and few working scientists. Science is just not at the table in a world where so many of the problems we face... fundamentally involve science.
>
> Working in Washington, it was so annoying to me to hear scientists so often complain about the low level of science literacy in this country;

> but then when we would ask "Can you help us out with this initiative or campaign?" they would go running and hiding in their lab.
>
> If we want to change the center of gravity in this country about public understanding of science and about issues like evolution or stem cells, scientists need to get personally involved. To just stand on the sidelines and carp goes against not only the best ideals of science but of democracy.

You can contribute to the political process because as a researcher you bring a unique and valuable way of thinking, asserted William Wulf and Anita Jones in an editorial in *Science*:

> Scientists and engineers think about problems differently. For example, lawyers, who disproportionately populate government positions, are trained to marshal an argument to support a predetermined conclusion (e.g., the client is innocent). In contrast, scientists and engineers are taught to analyze and design so that the outcome is not predetermined but is derived from the constraints of the problem. They collect relevant information, and only solutions that fit the data are acceptable. Scientists and engineers also think in terms of the total problem—for today and for tomorrow. An engineer will design a bridge to be taken down cost-effectively at the end of its life. This culture of thought and analytic tools and decision-making methods needs to have a stronger influence in decisions made about issues that at their root involve science or technology.

Wulf and Jones warn that "without sound technical input, some bad public policy will result. Without unrelenting oversight by individuals with technical expertise to ensure sound implementation, foolish actions will be taken."

Even if you are a junior researcher, you can consider involving yourself in policy issues, says Lehigh University's Sharon Friedman. "There was for a long time an unofficial rule that you don't talk about policy when you are young and untenured," she says. "You wait until you are a full professor, and you are fully established, and you are well known, and then you can start dealing with policy issues." However, says Friedman, that rule has all but evaporated.

Scientists also pull considerable weight with legislators, said David Goldston, former chief of staff for the U.S. House of Representatives Committee on Science. "Scientists often feel 'Nobody listens to us, no one respects us, we are not on the news every night,'" he told a 2008 AAAS meeting symposium on communicating science to the public. "Actually, scientists come in with a huge amount of credibility and deference given to them, and the question is how to use that yourself and how to make sure it is not misused by others," said Goldston, whose "Party of One" column in *Nature* covers science policy.

Scientists and their research findings are particularly valued by politicians for the political "cover" they provide, wrote David Murray, Joel Schwartz, and Robert S. Lichter in their book *It Ain't Necessarily So: How Media Make and Unmake the Scientific Picture of Reality*: "Since science is regarded as an objective picture of how things inexorably are in nature, data have the effect of absolving politicians of responsibility for a decision.... When challenged, the politician can retreat behind a screen of data and argue that science practically made him or her do it."

Work with Your Allies

Fortunately, there is considerable expertise at your disposal if you decide to involve yourself in legislation or public policy. You can find government relations experts in your own institution, in your scientific society, and in research advocacy groups.

First, contact your institution's own government relations officers. They are well versed on the policy and legislative issues your institution faces; and just as importantly, with the personal and political vagaries of those issues. They know how best to approach your own congressional members on an issue; how the history of that issue affects current political realities; and where you can best focus your energies to make a difference.

They can also help you develop effective communications, such as a letter to your member of Congress or congressional testimony. If you work for a government agency or federal laboratory, your communications with legislators must be handled especially delicately, and you should not consider any action without consulting with your government affairs office. While government relations officers know the machinery of government and the details of policies, you know better the science behind those policies. So your insights and explanations can greatly help them defend your budget or block unwise legislation that will affect you.

Your public information officer (PIO) might also help you develop communications for legislators, both national and state. University of Wisconsin PIO Terry Devitt recalls how he briefed the university's lobbyists on the negative impact of a state bill to criminalize human cloning or somatic cell nuclear transfer: "It was critical that legislation not be enacted, because of the perception that it would convey to the rest of the world about Wisconsin and doing science in Wisconsin," says Devitt. Such a message, he says, would damage the state's ability to attract high tech companies. "The risk of sending the wrong message or creating the perception that Wisconsin is not a good place to do science was very real," he says. "The lobbyists don't know much about stem cells or how science works," says Devitt. "My role was to help them appreciate the science, and what we are going to accomplish, and why it is important."

Your credibility as a researcher gives you particular authority in portraying the broader impacts of a piece of legislation, which greatly helps support the arguments for or against it. The Wisconsin stem cell legislation represents a good example of how PIOs and researchers can communicate such broader impacts. "We had to convey to the lobbyists, who in turn had to convey to the legislators, that—although they may only see their action as stopping what they might see as an unethical research project—they also send a message to the rest of the world that Wisconsin is a bad place to do science," says Devitt.

Besides meeting with legislators and testifying before committees on bills, you can also educate legislators on issues by creating Web sites, booklets, videos, and other communications. The University of Michigan produced just such materials as part of a campaign to reverse the state's history of restrictive legislation on stem cell research. "Here we are one of the leading biomedical research institutions in the world, with some excellent stem cell people, and we are stuck," recalls Duke PIO Karl Bates, formerly at the University of Michigan. "We really want to get into the game, but we can't because of this legislation. It didn't help any to be singled out in *Nature* magazine as one of the two states in the country with the most hostile policies toward science."

"The university found plenty of polling data that showed that teaching people about stem cell research increased their approval of the research," says Bates. So, to educate both state legislators and the public, he worked with researchers, programmers, and designers to create a Web site on stem cells. It featured an animation of basic stem cell science which was also produced as a DVD that scientists could hand out when they met with legislators. Besides reaching out to legislators and the public, Bates and his colleagues made sure that advocacy groups for diseases such as juvenile diabetes and Parkinson's and Alzheimer's disease received copies of the materials. "Just doing press releases or just testifying on the Hill isn't going to cut it," says Bates. "We live in an age where people can watch video on their cell phones, and we as science communicators need to get into that information space."

The government relations offices at scientific societies can be of great help in advocating for your research field. For example, the online resources for this chapter at *ExplainingResearch.com* lists such offices at the AAAS, American Astronomical Society, American Chemical Society, American Geophysical Union, American Medical Association, American Physical Society, American Physiological Society, Association for Women in Science, Association of American Universities,

Federation of American Societies for Experimental Biology, National Science Teachers Association, and Society for Neuroscience. And advocacy groups that seek to enhance general support of research and science education can also offer help in working with legislators. For example, the online resource section lists contacts for the Coalition for National Science Funding, Council on Governmental Relations, Coalition for the Advancement of Medical Research, National Association for Biomedical Research, STEM Education Coalition, Task Force on the Future of American Innovation, and the Science Coalition.

Also, the resources section for this chapter lists topic-specific science advocacy groups that can help you influence legislation on those topics. Such groups include the National Audubon Society (habitat and species conservation), Center for Science in the Public Interest (nutrition and health), Environmental Defense and Friends of the Earth (environmental issues), Mars Society (Mars exploration), National Wildlife Federation (species preservation), Planetary Society (space exploration), and Union of Concerned Scientists (environmental, security, food, and scientific integrity). There are also advocacy-group blogs, including Ocean Champions (ocean conservation) and Space Politics and NASA Watch (both space research).

A Basic Guide to Action

Even the strongest advocates for research among congressional members are still busy politicians. So, you need to communicate with them efficiently and in a way that will benefit both of you. Here are some basic guidelines for that communication:

- **Identify the right people.** Major advocates for your work will be your local congressional member, so get to know them and their interests. They may have specific staffers who handle science, technology, and medicine issues. Also, get to know congressional member who are members or chairs of committees overseeing the budget for your funding agency. If your local member of Congress serves on such a committee, you can be especially effective as an advocate. Your government relations allies can help sort out the players and how best to reach them. Other important contacts will be the staff of the policy offices and science-related congressional committees, as discussed in the next section.
- **Prepare your research spiel.** Develop a concise explanation that includes not only the concepts of your work, but also its benefits—including economic, societal, and even inspirational. For example, NASA recognizes that the stunning images from the Hubble Space Telescope represent not

only important basic science, but also dramatic evidence of the nation's commitment to exploration and technical excellence.

- **Tell stories.** Politics and policy making are not just about cold hard facts; they are a human enterprise, so human stories resonate with members of Congress and their staffers. They will respond to compelling case histories about discoveries in your laboratory or your field and their benefits. And these benefits need not just be lives saved or profitable products invented. They can also be inspirational stories of basic discoveries that broaden the intellectual horizon.
- **Share the glory.** "Reflected glory drives legislators," says science communicator Rick Borchelt. "At some level, they do believe in science as a fundamental American endeavor, but reflected glory drives them." Since reflected glory helps them get reelected, any approach to legislators "has to fit into their strategic goals for reelection or election to a higher post or into some other political end game for them," he says. Again, your government relations allies will know how to share the glory with legislators.
- **Help them get media coverage.** The most important way to share the glory is to help members of Congress get coverage in the local newspaper or on local television news. For example, you can help the congressional member get local media coverage of major federal grants. Even though they may have nothing to do with a grant, congressional members often announce such grants. If you have a major new grant, contact your congressional offices to ask whether they would like a translation of your project into lay language for their announcement. Other ways to garner media coverage for the congressional member include inviting local media to cover a member's visit or arranging for the news office to distribute a photo of the visit to local newspapers.
- **Establish a continuing relationship.** Among legislators' great annoyances are researchers who show up only when their funding is endangered, to argue for their budgets. Instead, establish a continuing relationship with your legislator, which can mean little more than dropping the congressional member or his or her staff occasional e-mails about your work, or sending them clippings of articles about your research achievements.
- **Send kudos, not just complaints.** Members of Congress love to receive thank-you e-mails on a vote you liked, or congratulations on passage of a helpful bill they aided. Take a minute to send such e-mails, reiterating an offer to help in any way you can. Consciously include in the message a laudatory quote they can use with their constituencies. Again, your

government relations allies can guide you on making such messages effective.

- **Do not lecture.** You may be a brilliant teacher, but legislators are not your pupils. Scientists "want to come to Congress and give tutorials," says retired representative Sherwood Boehlert, a longtime congressional science advocate. "That doesn't work. We don't have time for tutorials. They need to get right to the point: [telling the legislator] 'This is why it's important. I know there are a lot of competing interests, but here's why we should be at the head of the line. And here's what it means for society.'"
- **Do not whine.** When you meet with congressional members to talk funding, "Don't whine," says Don Gibbons, formerly at Harvard and now communications officer at the California Institute for Regenerative Medicine. "This is not about you and your troubles in your lab. This is about slowing down a cure for that congressperson's constituents. It is about how your problems can impact that congressperson and their constituents, so don't frame the issue as 'poor miserable me.'"
- **Make personal appearances.** Work with your government relations office to schedule a visit with the appropriate congressional members when it is strategically useful. Or, organize briefings on research-related topics. Such briefings can have a critical impact on the perception of your work by legislators and their staffs. For example, Gibbons cites the startling result of a survey showing that 40 percent of congressional staffers believed that most NIH funding stayed at NIH, rather than supporting research at universities nationwide. Gibbons—then chair of the National Communication Campaign of the Association of American Medical Colleges—and his colleagues launched a series of briefings to counter that misperception. "We would bring a university scientist and an NIH researcher together on a topic of a specific disease and really home in on the collaboration of NIH and universities and teaching hospitals," he says. "Two years later we had a retest, and we had a ten point improvement, where ten percent fewer people thought funding stayed in Bethesda; ten percent more knew that it comes from institutions."
- **Immerse them in your research.** Letting members of Congress see your research is the most memorable way to impress them with its importance. Schedule a visit for key legislators to your laboratory or center. In an interview with *Science*, Boehlert recalled the impact of visits by legislators to NSF research facilities in Antarctica and Australia:

> I was part of a bipartisan group of ten members. Of that ten, there were probably two who shared my view that global climate change was real and that we damn well better do something about it. The rest were skeptical or neutral. But after we got back, every one of them had a heightened interest in the subject. Why? Because down at the South Pole, they heard from scientists about how their experiments related to global climate change. The same thing happened at the Great Barrier Reef in Australia, where we heard how this great treasure was being damaged because of something called global [climate] change. And the next time there's a floor vote on the budget of some science agency supporting research on climate change…I'll bet that this group will be a more receptive audience because they've seen it firsthand.

Legislators' Hot Buttons and Levers

To figure out how to effectively pitch your issue to legislators, treat them like your favorite research instrument—that is, you create a specific input to elicit a specific output. Your aim in communicating with members of Congress is to elicit the output of supporting your cause. Here are the buttons and levers that control legislators' output:

- **Frame issues in terms of voters' interests.** Congressional members respond exquisitely to voters, because they work constantly to be reelected. Thus, your cause must resonate particularly with those constituents who will vote for that legislator. And traditionally, those constituents have cared about five categories of issues: economy, security, education, environment, and freedom/values, according to Francis Slakey, who is associate director of public affairs for the American Physical Society and Upjohn Professor of Science and Public Policy at Georgetown University. So, if you want attention for your issue, make your pitch to your member of Congress and staff in terms of how your research enhances one or more of those categories, Slakey told a workshop on working with Congress at the 2008 AAAS meeting.
- **Make your pitch concise.** Members of Congress receive a daily cascade of inputs such as contacts from constituents, so you will have only minutes to explain your position in your phone call, letter, or visit. Be able to do it accurately in no more than a few hundred memorable words.
- **Organize your pitch effectively.** Begin by asking for something, said Slakey. Do not preface your remarks with background. The congressional or staff member wants to know right up front what you want the

legislator to do. After you make your request, identify the problem and convey some sense of urgency, said Slakey. "If Congress doesn't have to solve a problem now, they won't," he said. "If it's a problem that will wait a year, they will wait two."

- **Complete your pitch by describing the solution, attaching it to one of constituents' five main concerns.** Says Representative Boehlert, "To talk about some great advance in pure scientific terms isn't enough. What does it do to strengthen the economy, or enhance competitiveness, or provide more jobs?"

Goldston cautions that you should also carefully separate the scientific from the political. "Think about whether what you are saying is a matter of science or whether it is a matter of policy," he told the AAAS workshop. "And try not to fall into the trap of conflating them, which both scientists and policymakers constantly do, both consciously and unconsciously," he said. For example, said Goldston, "the issue of whether climate change is real and human-induced, that is a science question. The question of what we should do about it; that is something where science should be a factor... but is not determinative all by itself."

"I am not suggesting that scientists shouldn't also engage on the policy side and express their own values," he said. "And that can even be part of the same conversation that you are having about the science... but you do want to be clear on when you are speaking as a scientist and when you are speaking as a citizen taking a policy position." Also, do not wrap yourself in your science, said Goldston. "Don't assume that anybody who has the same scientific information you do will reach the same policy conclusion. One of the great fallacies that all human beings have—but scientists especially—is 'If you knew what I knew you would think like I think.' And, you have to watch out for that."

Work with Policy Staff

Besides legislators and their staff, your other key contacts in Washington are the staff of the major federal policy-making offices. These include the Office of Management and Budget (OMB), the Office of Science and Technology Policy, and the congressional committees related to science and technology. Their Web sites are listed in the online resource section. While Senators and Representatives constitute the political arm of federal policy, these offices handle the nuts-and-bolts machinery of policy support for both the executive and legislative branches.

Their staff members range from scientists to lawyers, so you will need to understand to whom you are talking and adjust the technical level of your presentation accordingly. For example, you can expect more technical questions about your field from a lower-level policy staffer, but a higher-level policy analyst may want to know how your field fits into a bigger societal picture. However, you will still need to explain your work at a relatively nontechnical level even to scientist-staffers. "Any of us in science policy, regardless of where we are, have enormously broad portfolios, and we are operating far outside of our range of expertise by any stretch of the imagination," said Michael Holland, program manager in the OMB at the AAAS meeting workshop.

Prepare yourself before you go into a meeting with a staffer by understanding the exact duties and purview of the staff member and his or her office, as well as the issues of the moment, said Holland. Such preparation will avoid your embarrassment and save time. For example, you will not waste your time or the staff member's by talking about the NIH budget when the staffer works with the NSF budget. Begin by exploring the Web site of the office or committee—including the appropriation bills, reports, and hearings. And of course take advantage of the expertise of your government relations allies, as discussed previously.

In any encounter with a policy staffer, understand what is expected of you, said Holland. "Are you coming in as a technical expert who wants to inform somebody of the scientific underpinnings or implications of a policy question? Are you coming in as somebody representing your community?" As discussed above, come prepared to tell stories about your research, but with a caution, says Holland: "Make sure that when you put together a story that you also try to rip it apart...from both the technical aspects and the cynical aspects. You will meet a lot of staffers who will very happily take a meeting with you, and they have another agenda, and you may be feeding them tons of information that is going to let them shred what you care about or provide an argument against what you think your interests are." Also, always seek positive answers, emphasized Holland: "Never ever, ever ask a question in D.C. if you think the answer will be 'no,'" he said. "Once it is 'no,' we almost can't ever change it." When a policy staffer changes an answer, said Holland, opponents will accuse them of flip-flopping or being "squishy." Holland advises that, when you see a 'no' coming, propose that both you and the staffer think about the issue further and/or gather more information.

Finally, with staffers as with legislators, "focus on building a relationship over the long term," said Holland. "I never throw out a business card. I every once in awhile go squirreling around in my little pile of cards," for an answer to a question, he said. If you cultivate a long-term, credible relationship with a staffer critical to your field, you are more likely to have your point of view become part of the policy important to that field.

Warning: Your Research May "Mutate"

The political process will invariably "mutate" your findings as they are translated into policy, warn Murray, Schwartz, and Lichter in *It Ain't Necessarily So*. They describe an "almost alchemical" transformation:

> As laboratory results make it into the headlines, scientific uncertainty mutates into journalistic conviction. A train of events like this often unfolds in the following manner. First, a scientific body like the National Academy of Sciences convenes a panel to assess the state of our knowledge in an important area. The panel will likely be divided, as it evaluates a multiplicity of claims and demonstrations of varying quality. Out of the panel will come a recommendation, often rather tepid and sometimes even formally contested by the dissenting opinion of some panelists. A statement of this sort is then issued: "While no study is itself fully convincing, a pattern of evidence suggests that so and so is not unreasonable." But in a short while the qualifiers begin to drop out of the statement, and interested parties start to declare that "a preponderance of evidence now demonstrates so and so." By the time the investigation is deployed by a policymaker to justify a decision, all tentativeness disappears, and the media are told that "a substantial body of meticulous science has proven that x does in fact mean y." All of this ratcheting-up takes place without a single aspect of the underlying research having become more clear during the intervening period. But in the policy arena data are judged less by whether they are true or false than by whether they are useful or not.

If you involve yourself in the political process, expect your findings to mutate from tentativeness into certainty—which in truth may be the work of scientists as well as politicians. By understanding the mechanism of such mutation, you can learn to cope with it.

PART IV

Explaining Your Research through the Media

20

Parse Publicity's Pros and Cons

You are likely savvy enough about talking to a lay audience not to blithely spout equations to reporters, like the naive scientist in the cartoon that begins this section. However, besides mastering lay language, you also must learn how to use the limited bandwidth of the media to tell your research story. After all, the typical newspaper story is only a few hundred words, and the typical television news segment a minute or so. And those media stories are more and more likely to be done by reporters poorly educated in even the basic concepts of science and engineering.

You may wonder why the section on media relations comes last in this book, when the popular media would seem the most important outlet for your research explanations. The reason is that, realistically, for many scientists popular media are *not* their most important audience. Certainly, you would be enormously gratified to see an article on your work in the *New York Times* or a segment on the *CBS Nightly News*. But for most scientists, such a "hit" is unlikely—given the negligible capacity of popular media for research news.

So, why even bother reading this section, if you will not make *NYT* or CBS news? Because the techniques it teaches also will help you work with journalists from *American Scientist*, *Chemical & Engineering News*, *Discover*, *Geotimes*, *IEEE Spectrum*, *Nature*, *New Scientist*, *Physics Today*, *Science*, *Science News*, *Scientific American*, and the raft of other important science media. They may well be far more important to advancing your work than even the *NYT*. Working with the media presents both important benefits and pitfalls. Understanding these will help you accentuate the positives and minimize the negatives of media coverage.

Benefits of Working with the Media

Credibly Reach Your Audiences

Media articles offer corroboration of the significance of your work, read by your most important audiences. Those audiences will see that an independent journalist has chosen to cover your work, and the story likely contains comments on its significance from independent authorities in your field. This is not to minimize the credibility of news releases and in-house features about your work. If they are done well, they also carry weight with audiences. After all, they reflect a decision by your institution to highlight your work.

Discover Collaborators

Some of the most significant responses to publicity about your work will come not from the public, but from fellow scientists. And these scientists include those outside your field, some of whom constitute useful contacts or even collaborators. "We fairly often heard from researchers that they picked up good collaborators because of an article," says Julie Miller of her experience as a magazine editor. "For example, we'll report on work in botany, and a zoologist would read the article and recognize that there are shared interests and call up the researcher."

Spark New Ideas

It is not unusual for interactions with reporters to stimulate new ideas in researchers, says journalist Cristine Russell, a senior fellow at Harvard's Belfer Center for Science and International Affairs. "More than ever, I hear people thanking the news media and saying that the interesting questions that the reporters ask them stimulate them to think 'Well why aren't I doing that study?' or 'How could I do that research better?' And they really like the interaction and they are finding it enjoyable."

Reporters are particularly good at sparking ideas, because they tend to challenge you with provocative questions. They make you go back to first principles and distill your ideas down to their essence and in the clearest possible terms.

Give You Free Communication Tutoring

Your interaction with journalists gives you free tutoring on how to work with media and how to communicate. Your journalist contacts will no doubt be happy to share instructive tales of triumph and of horror covering scientists. A candid journalist may also alert you to your own weaknesses as a communicator. And just hanging out with a journalist can teach you communication skills by osmosis, says Chris Brodie, who is a neurobiologist-turned-research communicator and journalism teacher. "You're sitting there drinking beer with a guy who writes every single day and is under this incredible pressure to be clear and concise and to get to the point... the scientist could benefit from picking up a few of those tips," he says.

Help Your Field

Included in the readers of that newspaper article or Web feature on your findings are foundation officers, donors, and legislators who decide about funding your field. The news article on your work adds yet another tattoo to the steady drumbeat of successes that persuade those funders that your field is worth supporting.

Also, that story on your work might help correct a deficiency in media coverage of your field. For example, Lehigh University science communication professor Sharon Friedman cites the imbalance of coverage of the risks of nanotechnology. "In six years of coverage that we analyzed—which covered 38 newspapers in the U.S. and U.K. and two wire services—there were only 166 articles on potential environmental and health risks from this new technology, versus the thousands of articles about the gee-whiz wonders of nanotech," she says. "The field is being hyped ad nauseum."

Protect Yourself

An unavoidable reality is that if journalists become interested in your work, they will write about it with or without your help. The more you actively participate—explaining your work the way you want it explained—the more likely the story will be accurate and complete. Also, as discussed in chapter 9 on news releases, explicitly explaining your work's implications and giving proper credit protects you from charges of misrepresenting your work or hogging the spotlight.

Educate the Public

Media stories can help bridge the information gulf between scientists and the public. You might recall the previously cited Harris poll statistics showing the public's general ignorance about concepts like Earth's annual orbit around the sun, and the public's widespread belief in ghosts, UFOs, and witches. True, a few articles on your work might not totally correct such ignorance or alter widespread beliefs, but every word written about good science helps.

Possible Pitfalls of Working with the Media

Distract, and Even Detract, from Your Career

You might worry that a news story on your work will bring hordes of reporters to your doorstep, but that is not usually the case. Even the most dramatic research findings usually elicit heavy media response for a couple of weeks at most. If many journalists do call, just resign yourself to devoting the necessary time to interviews for that week. You can alleviate this crush by providing complete information on your Web site and by enlisting your PIO to manage calls.

Rarely, involvement in a hot controversy or a particularly sexy research story will ensnare you in a "hyperstory" that feeds on itself ad infinitum. The best course is still not to avoid the media, but rather to work with your PIO to provide information in the most effective way possible. Chapter 25 on avoiding communications traps covers how to prepare for a hyperstory.

Another pitfall is that you might enjoy the warming glow of the media spotlight a little too much, unduly influencing your career, points out journalist/author Keay Davidson. "I have seen too many instances where a scientist ends up getting in the news and getting a little bit addicted to it," he says. So, Davidson advises scientists to "think before they go for the spotlight, because it will distract from their research in all likelihood, and I worry it will distort their research agenda." You might also find yourself embarrassed by things that you allow journalists to persuade you to do, says Davidson. "I still recall how hurt and angry [Nobel Prize–winning physicist George] Smoot became when someone in his lab pasted on the wall a *People* magazine photo of him riding on rollerblades—which amusingly, the magazine had bought for him because he'd never done it before and they insisted on showing him doing something 'fun,'" he says.

More likely, media coverage will not negatively affect your research career, and you will worry too much. *New York Times* science reporter Sandra Blakeslee recalls one such instance of a scientist's overreaction. She had interviewed psychologist Jonathan Haidt for an article on how the brain handles moral dilemmas. In discussing how culture affects moral judgments, Haidt cited the example that in some cultures touching a menstruating woman is considered disgusting, coloring moral attitudes about such women. "I quoted him on the example, and he was mortified," recalls Blakeslee. "He wrote me a letter saying how upset he was. So, I wrote him a letter back saying 'I checked my notes and it was what you said, and that if I hurt you in any way I am so sorry.' I saw him a few years later and told him I was still sorry I had upset him, and he said 'Not to worry, not to worry, I was just at that point a young assistant professor and I was worried the quote would cast a bad light on me. And it had no bad effect at all.'"

Promulgate Errors

Perhaps your most nagging worry might be that media stories on your work will be rife with errors. True, print and television reports will invariably contain misquotations, misinterpretations, and factual errors. However, "rife" is in the eye of the beholder. What many scientists see as lethal flaws in a story may not be all that lethal. "If people are not happy [about media stories on their work] they should look to themselves and [understand] that their expectations might be too high," says Duke engineer/author Henry Petroski. "Newspapers only have so much space on the page, and journalism is a different medium than the technical paper."

Also, researchers may be holding media stories to the "wrong" standard of accuracy, points out Sharon Dunwoody, University of Wisconsin professor of journalism and mass communication. There are two kinds of accuracy, technical accuracy and communicative accuracy, she says. Technical accuracy measures the factual agreement between the scientific paper or interview with a scientist and the media story on the research. Communicative accuracy measures what the audience gleans from the story. Dunwoody holds that media stories should be judged not by technical accuracy, but by their communicative accuracy. To illustrate the role of communicative accuracy, she cites studies in which laypeople were asked to read and summarize a science article, after which scientists reviewed the article and the summaries. While the scientists found many technical errors in the original article, they nevertheless found a high accuracy in the audience's summaries of the stories. "The issue is, what is the fit between what the reader or viewer got out of this story and what the scientist said?" says Dunwoody. "Some studies suggest that what scientists are often talking about [when they measure accuracy] are details, like having all the researchers names listed; so if that is not happening then it is not accurate. But they are not talking often about the big picture—that when people read those stories, it is not those details that are important to them, anyway; it's the message. Obviously it is better for everybody if you get the little things right; but in the big scheme of things—regarding people's better understanding of the issues—they are not going to remember details, anyway," says Dunwoody.

Bottom line: you do not have to hold media stories to the same level of accuracy as, say, a technical paper, because the basic quality of communication with a lay audience will be perfectly accurate for your purposes. In fact, those errors in media stories that may bother you so much are nearly always far less damaging than the good those stories do your work and your field. And chapter 25 outlines steps to minimize such damage.

Create Envy

Publicity about your work might generate some envy among colleagues whose work, rightly or wrongly, receives less media attention. However, those same colleagues would likely also be just as green with envy at your successful grant, your seminal paper, or your prime laboratory space. You would not worry about that kind of envy. Why should you worry about envy of publicity that is just as integral to the success of your research?

Reveal Painful Truths

News stories may well bring out problems or situations that you do not particularly want aired. If there is such a possibility, consider whether correcting the problem rather than avoiding the media is more advantageous in the long run. Cristine Russell recalls just such an incident early in her journalistic career. During congressional hearings on smoking's link to cancer, "I was kind of shocked when the cancer institute officials were smoking in the hallway," she recalls. "I thought 'How common is this?'" So, Russell and an assistant called the offices of the National Cancer Institute and the public health agencies and asked who smoked and who did not. "We published this huge chart with names with little symbols of cigarettes and cigars and pipes." And, recalls Russell, although many of those listed were angry at the story, "one public information office of the National Cancer Institute said that the story was so embarrassing to him that he quit, and he thanked me."

21

Understand Journalists

Although you might think journalists are exotic, perhaps even threatening creatures, they are fairly easy to understand. For one thing, you as a researcher actually share many qualities with journalists, although there are major differences, says veteran science editor Tim Radford. As he explained in the book *Connections: Museums and the Public Understanding of Current Research*:

> Journalists and scientists have a lot in common: both are driven by curiosity; both regard the phrase "I don't know" as an interesting starting point, not an admission of defeat; both frame hypotheses, do literature searches, systematically gather evidence, write their results, and submit their articles for peer review before publication. There the likeness ends. Scientists take as long as they need to complete a paper; daily newspapermen do what they have to do inside a day. But that is not the important difference. A scientific publication matters even if hardly anybody reads it: it exists as a marker, as a record to be accepted or challenged, as a claim to priority. A newspaper story that was read by nobody would have been a complete waste of time.
>
> Scientists and journalists are, at bottom, looking for two different things. Both are concerned to find the truth. But the scientist wants an answer, however dull. The journalist would rather find a story. Both

> findings have to withstand the test of time, but the time in each case is different. That is why scientists spend five years or five months on a complex and profound piece of research that then takes five weeks to write up and another five months to finesse through the editorial board of a learned journal. And then journalists come along the next morning, ring them up, and spend five minutes asking them what the hell it means. What journalists write goes into a newspaper five hours later, and the next morning a reader picks it up, comes across a term like mitochondrion or functional genomics, and then stops reading in a fifth of a second, to go on to something else, perhaps something enjoyably disgraceful involving a politician and a call girl or a famous footballer and a fracas with the police.

Besides these similarities and differences, here are the other characteristics of journalists it will be helpful for you to understand.

They Come in a Broad Spectrum

Journalists are not a homogeneous group, but vary widely in expertise and ability. You will encounter journalists from *Science* or *Scientific American* who have PhD degrees and decades of experience covering your area. And you will encounter local newspaper reporters with freshly minted journalism degrees and a couple of general college science courses under their belt. Importantly, there is also a difference between science writers and "scientific" or technical writers, which many researchers miss. While science writers are journalists who seek to explain research to lay audiences, the job of scientific or technical writers is to craft clear technical prose for professional audiences. So, do not expect a science writer to be able to produce detailed technical articles, or a scientific writer to produce lay-level articles that engage the public. The chapters in this section cover how to effectively serve science and medical journalists.

They Work for Their Readers and Viewers

Besides not appreciating the nature of journalists, researchers also often harbor misconceptions about their role, says *Science News* editor Tom Siegfried:

> They think the role of the media is to promote science, or to put science in a good light, or to educate people about science, so science can be better off. That always leads to a little disconnect in the purpose of the

> interview.... The purpose of the media and the science journalist is to tell the public what scientists are doing, and why it is important or interesting; or what is wrong with it as well as what is right with it.
>
> To the extent that science does good things and is valuable for society, the coverage should reflect favorably. But to the extent that scientists do bad things erroneous things, fraudulent things...those things also are a part of what has to be reported.

Journalist/author Paul Raeburn says that, given the reportorial objective of journalists, dealing with them is straightforward:

> My general philosophy as a reporter is, I find things out, and I tell people what I know. If we keep it to that, it is quite simple, nobody gets confused, everything is out in the open, everything is clear; there is no ambiguity. My allegiance is to my readers. That is what I am interested in. I am not interested in any scientist's agenda or institution's agenda. I am trying to tell the readers something that they find interesting and entertaining.

Thus, the chapters that follow aim to help you give reporters interesting, entertaining, and accurate stories, while avoiding communications pitfalls.

They Work in a Pressure Cooker

The journalist who visits your laboratory returns to a pressure-cooker office. First, the journalist faces the pressure of gathering news and putting it out. "A newspaper is an accident that comes out every day," says science writer Robert Cooke. "You have no clue what is going to be in the next day's papers because it is all surprises."

The science or medical journalist faces the particular challenge of having to clearly explain complex subjects in every story—whether it is about a black hole, a cancer cell, or the earth's mantle. Imagine if sports reporters had to explain in each story why three strikes means the batter is out. Or, imagine if the crime reporter had to explain what a bank robbery was; or a political reporter had to explain the basics of campaigning.

Science and medical journalists also face the pressure to learn new subjects every day—as if sports reporters were shunted over to covering crime, and vice versa. Says Cooke, "Every story is different, and what you know from the previous story does not count. It doesn't fit anymore if I have done a story on geology, and it's geochemistry instead. Not only are they different subjects, but all these subjects are moving, and the pace is picking up. What you knew yesterday may not be true today."

The demands of editors also place particular pressure on science writers. The journalist who interviewed you does not simply zip back to his or her newspaper, pound out a story and see it published. That journalist must go through an editor, a process that for science and technology stories is more fraught with problems than for general news. Cooke declares that editors at a daily newspaper

> consider science writing as just [covering] the crying babies of the local hospital. That is all. Anything beyond that, their eyes glaze over when you say science, and unless it is really hot stuff, they don't want to do it. Sometimes you have to hit them with a two-by-four to get their attention, because they are worried about cops and robbers, and school boards, and sports. And science does not rank high in there. It does not generate big headlines which is what they like.

What's more, says Cooke, reporters must deal with a constant turnover in editors. In 11 years at the *Boston Globe*, he had 14 different editors. "You are always breaking in a new editor who doesn't know anything about science or care." And, says Cooke, that editor is usually in charge of the "trained seals," which is the pejorative name given specialty reporters. "He is in charge of the religion beat, the education beat, the science beat, and the medical beat and doesn't know anything about any of them."

Besides these traditional pressures, journalists, like researchers, are under new pressure from an increasingly active audience, as discussed in chapter 1. Describing this new audience Bill Kovach and Tom Rosenstiel wrote in their book *The Elements of Journalism*:

> Technology is transforming citizens from passive consumers of news produced by professionals into active participants who can assemble their own journalism from disparate elements. As people Google for information, graze across a seemingly infinite array of outlets, and read blogs or write them, they are becoming their own editors, researchers, and even correspondents. What was called journalism is only part of the mix, and its role as intermediary and verifier, like the roles of other civic institutions, is weakening. We are witnessing the rise of a new and more active kind of American citizenship—with new responsibilities that are only beginning to be considered....
>
> Among other things, people have the ability to interact with the news itself as well as the professionals delivering it. Some use the Web to present their own accounts of events, complete with photographs,

> video, or audio. Some contact the journalists covering a story through e-mail or feedback forms to either correct the record or offer new facts. And some participate in discussions about the process that brought forth the news, building an almost immediate record of press criticism and scrutiny.

This active audience exerts increasing pressure on journalists, who must figure out how to respond to such criticism and scrutiny, while still responsibly reporting and verifying their information, wrote Kovach and Rosenstiel.

Besides transforming journalists' audience, technology is also transforming their duties. Besides writing for their traditional print outlet, they are now pressured to produce a constant flow of "instant news" for media Web sites, and to produce blogs and multimedia. A survey of science journalists by *Nature* found that, as the number of science journalists dwindles, 59 percent of those remaining have seen the number of items they work on increase. Juliet Eilperin, environmental reporter for the *Washington Post*, describes the effect of those pressures: "The pressure to constantly freshen your Web page is so immense that it is causing people to put things up before they are fully vetted. This is something that no mainstream news organization has been able to negotiate well, which is that there are different standards for the Web and for the print edition...and it still has the names of our publications on it, and our name is on it."

Declares *New York Times* science writer Cornelia Dean, author of the book *Am I Making Myself Clear?*, "The one and only asset we have in this environment is our franchise; it is the brand name and absolutely anything that diminishes the brand name, in my opinion, is lethal. But there is all this pressure for content, content, content."

Pressures to feed the Web are also requiring journalists to become combination reporters/bloggers/multimedia producers, says *New York Times* environmental reporter Andrew Revkin, leaving less time to research and reflect. "The Web is essentially filling all that space with more production, so you are actually getting more production, more words....I write a thousand words a day for the blog, no matter what I am writing for print."

They May Be an Endangered Species

The cadre of trained, experienced science writers in popular media is dwindling. One reason for the decline is that most newspaper and general magazine editors see science writers as peripheral to the main news mission. Editors also see science writers as "expensive," in that their stories take more time and expertise than do those of general reporters or even those in some specialties such as sports. Thus, newspapers have steadily reduced their science writing staffs, and those reporters who remain may not really write about research, but about health and medicine.

More generally, says Dean, "I see the mainstream media melting down and losing resources. There are always going to be people who are going to be prepared to pay for high quality, reliable information. The thing is we don't know at this moment what that landscape is going to look like in five or ten years, because it is changing completely." Scientists need to understand how these pressures affect science reporting, says Dean; "somebody who is the one reporter on the paper, who yesterday was at the zoning board, and this morning was at the auto wreck, and now in the afternoon is at the university laboratory—the fact that person is not absolutely up to speed on whatever is going on in your subspecialty is not an indication of sloth."

They Are "Journal-ists"

The pressures to produce and to simplify their stories have transformed science journalists into "journal-ists," says Ben Patrusky, executive director of the Council for the Advancement of Science Writing. That is, they concentrate on covering journal articles, which reflect the individual, sometimes evanescent, puzzle pieces of science, rather than assembling those discoveries into broader overviews of a topic. "I know we understand—those who do science writing—that just because it is a journal article doesn't mean it's true," says Patrusky. "It is an observation for the moment which is going to be tested.... But somehow because it has been in a journal like *Science*, it takes on the mantle of truth." Thus, he says, reporters do not and, in fact cannot, do "enterprise reporting," in which they cover the uncertainties, contradictions, controversies, and complexities of a topic.

To science communicator Rick Borchelt, such a simplistic portrayal of science is "a narrative of hubris: it perpetuates the view that science is a linear process of steps and breakthroughs, and gives no account of the trials and errors that actually occur along the way," he said in an article, "How Journalism Can Hide the Truth about Science." What's more, says Siegfried, even covering journal articles has become more difficult:

> These journals started out with a handful, like *Science* and *Nature* and the *New England Journal of Medicine* that would promote what they were

> going to have coming out. But now you have dozens of journals that are...pushing out these releases saying "Our new paper shows this and this and this." [As a result] there is not enough critical judgment about some of these papers, and nobody looks deeply at the statistical evidence that is being used to reach this conclusion. They just get the conclusion.

Revkin decries such "whiplash journalism," in which reporters "cover every story today as if that is the new reality, whether it is about Greenland ice trends or...sea ice in the Arctic. If you don't put it in context with the larger trajectory of what we know or don't know about the ice or species loss, then you are really doing the reader a disservice; you are actually disengaging the public from the value of journalism."

They Stink at Statistics

In their stories, science writers generally do not explain the statistical methods that pertain to a study's validity. For example, they seldom point out that a study might be weak because it depends on indirect measurement of behaviors reported by subjects, rather than by direct measurement of those behaviors. Nor do science writers usually distinguish between association and causality, often assuming that a statistical link between two things proves that one causes another. An example of such statistical ignorance was the egregious media coverage of a survey in which women were asked questions about their cell phone use during pregnancy and their children's later behavior. The researchers who conducted the 2008 survey found a statistical relationship between higher cell phone use and more behavioral problems. Misreading these statistics, the British newspaper the *Independent* published an alarmist story headlined warning "Using a Mobile Phone While Pregnant Can Seriously Damage Your Baby." The story declared that "women who use mobile phones when pregnant are more likely to give birth to children with behavioural problems, according to authoritative research."

Discover magazine criticized the faulty coverage, pointing out that "correlation does not equal causation." The magazine noted that the linkage between cell phone use and children's problems might have been due to factors unrelated to cell phone use. For example, mothers with access to cell phones might spend less time with their children, noted the magazine.

Journalists' lack of familiarity with statistics also leads them to depend on news releases to highlight the significant findings, charge Murray, Schwartz, and Lichter in *It Ain't Necessarily So*. Such dependence means that journalists "can easily fail to understand the significance of data when that significance is not pointed out to them by researchers who compile and issue the data," they wrote.

Journalists also usually neglect to offer perspective on the overall risk that, for example, oil spills or forest fires represent. They focus instead on specific oil spills or forest fires, wrote the authors. "If the media seldom excel at quantifying risk, that is largely because doing so might get in the way of telling a good story (by which we mean an exciting, not necessarily an accurate) story related to risk.... An attention-grabbing rhetoric of risk—you'd better worry, because we're in danger—can sometimes trump the more prosaic reality that a risk is relatively inconsequential."

There are, however, effective antidotes to journalistic neglect of statistics. For example, countless journalists have learned to report statistics accurately from the classic book *News and Numbers: A Guide to Reporting Statistical Claims and Controversies in Health and Other Fields*, by Victor Cohn and Lewis Cope.

They Cover Stories "In-Shallow"

Even when science journalists do cover stories "in depth," they tend to simplify and dramatize those stories. They look for heroes and villains, winners and losers, criticism and controversy, and of course what is new and different about their topic. Thus, they seldom cover the complexity and process of science. As Murray, Schwartz, and Lichter wrote in *It Ain't Necessarily So*, "Interest in drama means that the qualifications, caveats, and uncertainties that are the bread and butter of scientific research can instead be treated as the roughage of journalistic accounts." However, scientists also share some of the blame for oversimplifying explanations of their research, charged the authors: "Researchers are more to blame than reporters for ignoring alternative explanations; if researchers offer only one cause, what are reporters supposed to do, other than accurately summarizing their findings?"

San Francisco Chronicle science writer David Perlman points out that, given the limitations on reporters, such shallow coverage is unavoidable. "First place, a story like that takes a lot of space," he says. "Second, for the reporter it takes a lot of time. Third place, editors are unlikely to be interested in it." Of course, there are exceptions to such "in-shallow" reporting—for example the many excellent popular science books, and the substantive feature stories in such magazines as the *New Yorker*, *Smithsonian*, and *Discover*.

They Accentuate the Positive, Eliminate the Negative (and Seldom Mess with Mr. In-Between)

Science journalists also tend to be Pollyannaish about research discoveries, due mainly to space limitations and editors' preferences. Says journalist/author Keay Davidson, editors

> want science stories, but they don't want any science in them. They don't want them to be technical, and they particularly don't want the stories to be anything less than upbeat. If you write about stem cells, it had better be the greatest thing since cream cheese. Any expression of doubts or uncertainties or ambiguities; any suggestion that people are overselling their findings, any suggestions there might be environmental issues involved; these things are very hard to get into print. Most editors will say, forget that, we want the sizzle, and to heck if it is a steak or not.

In accentuating the positive, pointed out Vincent Kiernan in his book *Embargoed Science*, "journalists are far more likely to report on a 'positive' research finding—such as the conclusion that a drug has an effect on patients—than [on] a negative finding of no effect."

Medical journalists also tend to emphasize the hope for cures from a scientific advance, rather than the cautions. In this "symbiotic relationship," wrote Daniel Greenberg in his book *Science, Money, and Politics*, "the science press tends to be in uncritical harmony with the people it writes about." He wrote that "popularizers of medical research frequently employ a set piece that declares substantial grounds for hope against a so-far intractable disease while cautioning against undue optimism."

They Are Two-Sided, but Only Two-Sided

Journalists all too often make it a practice to distinguish two sides of a story even if there are more than two sides. Stanford Climatologist Stephen Schneider describes the quandary facing scientists in his online essay "Mediarology: The Roles of Citizens, Journalists, and Scientists in Debunking Climate Change Myths":

> A climate scientist faced with a reporter locked into the "get both sides" mindset risks getting his or her views stuffed into one of two boxed storylines: "we're worried" or "it will all be OK." And sometimes, these two "boxes" are misrepresentative; a mainstream, well-established

> consensus may be "balanced" against the opposing views of a few extremists, and to the uninformed, each position seems equally credible. Any scientist wandering into the political arena and naively thinking "balanced" assessment is what all sides seek (or hear) had better learn fast how the advocacy system really functions.

Journalists may also cover both sides of a story, even if one side is scientifically groundless. For example, says Perlman, in the case of evolution, "if the intelligent design people come to town and have a public presentation, we will cover it. I would say I would cover it biases and all. I would give them a reasonable amount of space in my story, and I would know exactly who I would counter it with." However, emphasized Perlman, he and other experienced science journalists only write about intelligent design if its proponents produce news. "Any other story on evolution I don't go and ask intelligent design people to give them the ultimate possible excuse for being credible, when they have none."

They Limit Covering the Same Topics

When you read a newspaper story about an advance in your field, you might be tempted to contact the journalist to write another story featuring your work. However, such a story is unlikely because journalists and their editors purposely limit coverage of topics and institutions. "With so many topics in science today, we can't keep covering the same thing over and over again," says Julie Miller, who has edited *Science News* and *Bioscience*. "So sometimes we are in a situation where if we have done a story on breast cancer, and something even more important comes up the very next week, we just won't cover it in the same way. We may save it up and do it as a feature or we may do something shorter. It is not at all fair, but we just can't look like we are the magazine that only covers breast cancer."

They Need to Do Their Own Reporting

Even though your institution has issued an utterly brilliant release on your work, do not expect reputable reporters to use the release as is. "Scientists don't understand why [reporters] keep calling because they have to get something fresh," says Robert Cooke. "They don't want to see exactly the same quotes in every paper in the country. I will rewrite the news release and put my own spin on it, because if the paper is just running the news release they don't need me," he said. "You've got to get something...different than the wires and make your own editor sit up and take notice."

However, the practice of independent reporting is eroding, charged journalist Cristine Russell, a senior fellow at Harvard's Belfer Center for Science and International Affairs. She wrote in an article "Science Reporting by Press Release" in the online *Columbia Journalism Review* Observatory:

> A dirty little secret of journalism has always been the degree to which some reporters rely on press releases and public relations offices as sources for stories. But recent newsroom cutbacks and increased pressure to churn out online news have given publicity operations even greater prominence in science coverage.... In some cases the line between news story and press release has become so blurred that reporters are using direct quotes from press releases in their stories without acknowledging the source.

They Need a News Peg

Reporters need a reason to write a story at a particular time, unless it is a feature story. As discussed in chapter 10 on news releases, the best news peg is a scientific paper or symposium talk. Another news peg might be the publication of a formal report on your research finding to your funding agency. Without a news peg, your research finding will garner far less attention.

They Work on Tight Deadlines

You might not appreciate how much time is of the essence for reporters. After all, as Tim Radford points out in the quote at the beginning of this chapter, your work took years to accomplish, many months to publish, and its findings are timeless. However, tight newspaper deadlines mean you might have a window of only minutes to have your perspective portrayed in a story. Fail to call a reporter back immediately, and you might miss an opportunity to correct errors or to get your two cents' worth in about a rival's research finding. Even a weekly or monthly magazine might be on a tight deadline. "If people aren't there when you call, you don't wait for them to call back; you call someone else," says Cooke. "You have a deadline coming, and if you miss the deadline you are not in the paper." So, call a reporter back immediately, and also instruct whoever answers your phones to automatically ask reporters what their deadline is.

They Like Visuals, Multimedia

A stunning photo or compelling illustration can make the difference between your research being featured prominently or relegated to a lesser spot in a newspaper's print edition. Similarly, video or animations will attract more attention on a media Web site. "All newspapers are now basically Web sites, and we are now looking for, and I am looking for in every story...any kind of visuals," says Sandra Blakeslee of the *New York Times*. What's more, she says, media Web sites are far more likely than in the past to link to a researcher's Web site.

With all these characteristics and needs of journalists in mind, chapter 22 covers how to meet those needs, to get your research story out.

22

Meet Journalists' Needs

Even before you sit down for your first interview with a journalist, you can lay the groundwork for a productive relationship. The steps described in this chapter to lay that groundwork might seem only common sense and common courtesy. But all too often researchers neglect to take them. This failure not only reduces the likelihood that a reporter will cover your current paper satisfactorily; it can also damage your long-term relations with the reporter.

Be Willing and Available to Talk

Before you even consider issuing a news release or giving a public talk, commit to making yourself available to talk to journalists on the record. "It is terrible to get halfway through [an interview] and find out that they have reservations either about the data or about whether they really want this to be public," says editor Julie Miller. She recalls a particularly frustrating incident in which she received a news release and developed a story based on the discovery it described. "I called the researcher to check on a couple of facts...and he didn't want to talk with me, because he didn't want the story to come out at that point. He and the PR person should have figured this out ahead of time." Miller said she was particularly chagrined because other magazines and newspapers were doing the story off the news release, and she was being importuned not to write it because she took the time to check facts.

More commonly, researchers are perfectly willing to talk when a release is issued, but are not available due to travel or other exigencies. However in these days of international cell phones and worldwide e-mail access, there is no reason you cannot be available anywhere. But if you are truly not going to be available—on a desert island or the depths of a jungle—designate a spokesperson who will be accessible.

Give the Full Story

Do not try to be "a little bit pregnant," when giving a story to journalists. Giving only partial cooperation or an incomplete story is neither reasonable nor ethical. National Public Radio (NPR) science correspondent Joe Palca describes an egregious example of such incomplete cooperation. His "partial-pregnancy" story began in 2006, when a commercial PR rep ardently courted him to cover an upcoming clinical trial by the company StemCells, Inc. The company was testing a neural stem cell treatment for Batten disease—a fatal neurodegenerative disorder. Palca and an NPR colleague agreed to cover the clinical trial, planning to feature a family whose child suffered from the disease, and who had agreed to be interviewed.

But when the company held an audio teleconference, the PR person abruptly decided not to allow Palca to personally attend the teleconference to record broadcast-quality sound, or to interview the scientist or patient afterward to obtain the same quotes as in the teleconference. The PR rep said "We are only prepared to give this one statement, and you won't be hearing from us again," recalls Palca. The PR rep cited patient privacy as a reason not to cooperate. But Palca retorted that the patient and his family had already agreed to talk to NPR. The refusal to cooperate led to public embarrassment for the PR person and the scientists, as is often the case. Palca highlighted their refusal to cooperate in the NPR segment he ultimately produced.

Clear Bureaucratic Roadblocks

Make sure that a reporter does not have to cope with an internal bureaucracy when pursuing a story about your work. Sandra Blakeslee of the *New York Times* tells of the frustrations of such red tape. "One thing that drives me completely crazy is when you are in a real hurry, and you call up a researcher and say 'I just found your paper, and it's fantastic,'" she says. "And the researcher says 'Oh, I would like to talk to you, but you have to talk with our PIO [public information officer] first.' And then you have to make another phone call to reach that person, and they are out to lunch, and their message says to call a secretary, who can't give you permission. I will sometimes say, 'Well, screw it!'"

Certainly, government or corporate laboratories might reasonably require permission from a PIO for their researchers to talk to the media. However, unless your work involves significant issues of security or proprietary information, such permission should not be necessary. Sometimes your administrator or PIO may impose an inappropriate requirement for such permission arising from the administrator's need for political control or simply the PIO's need to feel involved in the process. On the other hand, nonintrusive involvement of a PIO can be invaluable in clearing the roadblocks for reporters, says Blakeslee. "They can help me with getting more information, getting me in touch with the researcher if he or she is traveling or it is after hours, getting pdfs of a paper, and getting graphics," she says.

What's more, a PIO's involvement can be useful in preparing both the reporter and you for an interview. Such PIO service is covered in the special online section at *ExplainingResearch.com* on working with PIOs. Even if your PIO's involvement is not required, however, it is a good idea to notify the PIO of interview requests.

Consider Communication Training

If you will face reporters more than occasionally, strongly consider communication training. Usually, such formal training consists of a one-day course, but even an informal coaching session with your PIO would be helpful. In the past, such courses were called "media training," but researchers are often uncomfortable with the implication that they are seeking to become publicity hounds. So, these days, the more comfortable labels are "presentation training" or "communication training."

In such workshops, you will usually practice giving television and/or print interviews, which are videotaped and your performance critiqued. You might be startled, even shocked, at how awkward and inarticulate you come across at first. But with experience, you will learn to use gestures, inflection, and lay-level explanations effectively. Such training also will help you refine your talking points and practice fielding tough questions. Skill at handling hostile questions can be a reputation-saver, as chapter 25 on avoiding communication traps illustrates.

Besides improving your interviewing technique, such training will benefit your professional communications, says Johns Hopkins PIO Joann Rodgers. "The very same techniques that you need to learn in crisis communication, you can use to tell a story to your colleagues at the next scientific conference," she says. "Or, what if you are arguing on behalf of a grant you want? How do you make your case? How do you find those essential elements...that are going to get people to understand your point of view?"

Your institution or professional society might offer communication workshops, or your news office might create a program. Another possibility is joining with other researchers to pool grant outreach funds, such as those given by NSF, to establish training workshops and hire an outreach coordinator. Some institutions offer full courses in lay-level communications, such as Cornell's comprehensive training class for graduate students. In a letter to *Science*, professor of science communication Bruce Lewenstein and colleagues who developed the course made these recommendations for such courses:

- Involve journalists, scientists with experience communicating their research, and media relations staff who "have a unique perspective on what topics are newsworthy and on the challenges scientists face in communicating effectively."
- Visit newsrooms to talk to reporters and sit in on editorial meetings. "This process reveals what stories interest reporters and how those stories are developed. Understanding this process will help scientists identify and explain the newsworthy attributes of their own research."
- Give hands-on experience, for example, writing press releases and articles, giving and getting interviews, creating a Web page, and setting up a science blog.

Even if you skip formal training, though, *Explaining Research* offers a thorough grounding in media techniques that will be helpful. Also, there are concise media guides, such as the American Geophysical Union's "You and the Media: A Researcher's Guide for Dealing Successfully with the News Media," by Herbert Funsten, which is available online through the references at *ExplainingResearch.com*. There are also good discipline-specific guides, such as *Communicating Astronomy with the Public 2007*, edited by Lars Lindberg Christensen and Ian Robson. And for PIOs, there is *The Hands-On Guide for Science Communicators: A Step-by-Step Approach to Public Outreach*, by Christensen.

Become a Credible Source

Helping journalists understand your field in general does both them and you a favor. You give them useful information, help ensure the accuracy of media coverage of your field, and engender mutual trust between you and journalists. As science communicator Lynne Friedmann puts it, "You need to become a resource to the press before you become a source. The first time you talk with a reporter shouldn't be the day that your big research paper comes out. You want to have that kind of context beforehand. Even though such calls may or may not lead to a story, it is not wasted time on your part. You start to develop a relationship with a reporter."

San Francisco Chronicle science writer David Perlman says that in his decades of covering science, such personal relationships have helped establish credibility. "There is a gradual build-up of trust—the fact that I have gotten to know them personally, and I have assessed them unconsciously as a straight-shooter kind of person. The more you personally get engaged with a scientist the more you feel you are able to trust them."

Becoming a source may only mean giving journalists your cell phone number and e-mail address, and inviting them to contact you if they need help on a story. Or, you might become more activist, alerting reporters to newsworthy work in your field, says *Science News* editor Tom Siegfried. "If you know a journalist is really into a certain aspect of astronomy, then when you find out about something in that area, you call or send an e-mail about a meeting or workshop coming up, and even send them an invitation," he says. "If you look at it as being helpful, as opposed to promoting your own work...that is a better foundation for the journalist to pay more attention to your work later on." More formally—as discussed in chapter 8 on forging your research communication strategy—you can volunteer to be on experts lists maintained by your professional association, AAAS's Science Talk and EurekAlert!, the Science Media Centre, and ProfNet.

Make Communication a Two-Way Street

The new interactivity between journalists and their readers gives you a golden opportunity to ensure that your field and your work are covered accurately. You can instantly comment on a journalist's story online or send the journalist an e-mail. Or more traditionally, you can send a letter to the editor or request a correction to a story. Far from being irritated by such communications, journalists and their editors see them as energizing their publications, says Miller: "We love to run letters that disagree with us or even say we made a mistake, because that means people are really taking us seriously," she says. "If we can run a letter from some scientist saying that he was upset because we said nanograms instead of milligrams, it means they are paying attention to us."

Pitch Story Ideas

Journalists also like to receive pitches on feature stories on your work, but only if they are well conceived. A useful story pitch ranks high on newness, importance, and interest, says Siegfried. He advises that story-pitchers search news sites such as Google News to see what stories have already been done. "Figure out what is a different angle, or covers the topic in a way that nobody has done before. Also, what are the elements that make the story newsworthy and inherently interesting to the audience that make them want to read the story?"

E-mail is the best way to deliver a pitch, since a journalist can more easily manage the information and pass it to an editor. The e-mail pitch should include a clear subject line and a concise explanation of the idea—including why it is new, important, and interesting. Make a pitch over the phone only if you already know the journalist and want to informally test an idea before putting time into developing the full pitch. While in some cases you might pitch a feature story to a journalist yourself—for example, if you know the journalist well—in most cases it is more effective and politic to involve your PIO in developing the story idea and making the pitch. So, start the process by suggesting the story idea to your PIO.

Make (Good) Things Happen to Journalists

News is not what happens; news is what *happens to journalists*. That is, you will get a better story to the extent that you involve journalists in your work, enabling them to make their stories more immediate and engaging. So, besides offering a journalist a standard interview, news release, feature article, and/or Web site URL, invite the journalist to trek into the field with you, tour your laboratory, and even participate in experiments—preferably nonhazardous ones. A story written from a hospital bed will likely not be favorable. Invite journalists to seminars that might interest them and even to your journal club. Of course, make sure that the seminars do not discuss work whose scientific publication might be compromised by a premature media story.

Also, where useful, give journalists "samples" of your work. For instance, when I worked at Cornell, we offered journalists samples of new foods developed by Cornell scientists—including canned green beans produced by a new process that kept them crispy. And when PIO Sue Nichols was at Michigan State University, she offered journalists packets of coffee that exemplified the university's project to help Rwanda and Burundi export specialty coffees. Imagine how receptive journalists were to doing a story on that coffee when the sample gave them their morning caffeine hit.

You can also involve journalists in your work by developing media workshops, to help them better understand your field. University of California–San Diego's "Molecules for the Media" press workshops represent an excellent example. Organized by PIO Kim McDonald and his colleagues, the workshops covered topics for which the university has a strong cadre of researchers. The topics also were those that journalists would likely have to write about—including atmospheric chemistry, molecular understanding of disease, intelligent nanosensors, and biofuels. Workshop panelists gave brief presentations, followed by a roundtable discussion, all of which was videotaped, broadcast on local cable television, posted on the workshop Web site, and produced on DVDs. "The idea is to bring them up to speed on a subject, not to necessarily have them do a story that day, but get them familiar with the experts that we have here, so they can put our scientists' names in their Rolodex," says McDonald. Such seminars need not be large events. They can be informal lunches or dinners with a researcher and a few reporters, at which the researcher discusses his or her work and field.

23

Prepare for Media Interviews

The interview for a news or feature article constitutes your most important encounter with a journalist. Preparing for the interview, as described in this chapter, will make it far more likely that the article on your work will be accurate and newsworthy.

Understand Your Interview Bill of Rights

When a journalist arrives at your door, the most important knowledge you need is the rights that you have, and do not have, in an interview. This knowledge could not only make you a more effective source for the journalist, but also save you from embarrassment or worse. Here is the list that Don Gibbons, communications officer at the California Institute for Regenerative Medicine, shares with researchers:

You have the right to

- Know to whom you are talking
- Know the show or publication
- Understand the focus of the story

- Know who else is being interviewed (especially if you are part of a television panel, with others being interviewed remotely)
- Be told how your point of view fits in
- Know the format of a radio or television segment—live, taped, or call-in
- Set a reasonable time limit on the interview
- Postpone the interview until you are ready (keeping in mind the reporter's deadline)
- Answer questions without interruption
- Correct your misstatements during the interview
- Use notes
- Record the interview

You do *not* have the right to

- Know questions in advance, although you may certainly ask what the reporter will be interested in knowing
- See the story before publication or broadcast, although you may offer to check facts and/or quotes
- Change your quotes, although public information officers (PIOs) routinely allow editing of quotes in news releases
- Edit the story
- Expect to be the only source
- Dictate publication date or placement

Decide Who Should Do the Talking

Before any interviews, designate who should talk to journalists and under what circumstances. Besides having a spokesperson for individual journal articles and symposium talks, also designate one or more for your laboratory as a whole. For journal articles, the spokesperson is usually the corresponding author, even though a junior first author might have done most of the work. Journalists prefer to have only one spokesperson for news stories about journal articles, so although you might want to highlight more than one author, it does not usually work. And, of course, for talks the spokesperson is the speaker, even though the talk may list more than one author. If a paper's authors are from different institutions, the main spokesperson is usually the senior author from the most central institution. If the collaboration was equal, decide which senior author will do the talking.

Local reporters will likely want to quote the local participant in the research, which is perfectly permissible. If a journalist decides to call an author who is not the spokesperson, that author should certainly be allowed and willing to talk to the journalist. Do not shut off a journalist simply because you are not the spokesperson, but feel free to suggest that the journalist also call the spokesperson. However, emphasizes Sandra Blakeslee of the *New York Times*, "All of these caveats go right out the door if somebody says 'My deadline is now.' They really have to respect that and be concise, and be quick, and be accurate."

Also designate a principal spokesperson for feature articles on a project, laboratory, or center. The default principal spokesperson is usually the most senior researcher or laboratory director. However, for a large laboratory whose work draws heavy media coverage, the director might designate another researcher to act as spokesperson. But as with news articles, if a journalist wants to talk to anybody else in the laboratory or center, those people should be made available for interviews.

Do Your Interview Homework

"An interview is no place to have an original thought," declares Johns Hopkins PIO Joann Rodgers. "You need to think about what you're going to say first. That doesn't mean you can't get creative and spontaneous with language. You might think of a new analogy, metaphor, or anecdote when you're talking, and that is fine. But you have to at least think through the story you want to tell before you sit down. It's like the old saying, if you have an hour to chop down a tree, spend fifty minutes sharpening the ax," she says.

Fortunately, sharpening your "interview ax" is not a complex process. It involves basically scouting out your journalistic audience and outlining what you want to say to them. Here are the steps to take:

Find Out to Whom You Are Talking

With only modest effort, you can find out about the reporters interviewing you, so you can effectively explain your work to them. First, understand their media outlet, which will help you decide what information the reporter needs, says science journalist Patrick Young. "If you are talking to somebody doing a news section piece for *Science* for instance, you can be a lot more scientific in your language than you will be if you are doing it for the *Washington Post*," he says.

A caution: make sure you *really* understand which news outlet you will be talking to. Science communicator Cathy Yarbrough recalls how lack of such understanding can cause embarrassment. "When I was at the Arthritis Foundation, scientists would talk to the *National Enquirer*, and when the piece came out, they would call me or my boss livid at being published there. So we would ask, 'How did they identify themselves?' and they would say 'Oh, the *Enquirer* newspaper.'" Of course, the scientists had confused the sensationalist tabloid *National Enquirer* with the newspaper the *Philadelphia Inquirer*.

Find out the reporter's level of understanding and expertise by asking your public information officer (PIO) and/or the reporter. The reporter will appreciate your reconnaissance, says *San Francisco Chronicle* science writer David Perlman. "It's helpful when a reporter walks into the lab of a senior researcher if the PIO tipped off the scientist that 'This guy you are really going to have to be very simple,' or 'This guy pretty well understands what you are doing already, so don't be afraid to explain it in more technical terms,'" he says.

You might even ask for references, especially if a reporter is planning a longer piece that will require you to invest considerable time, says journalist/author Jon Franklin. "One thing that always surprises me is that scientists never ask for references," he says. "It would be very easy to ask 'Who have you interviewed and who would vouch for you?' and also read something they had written. You would understand what kind of a person they are...and find out if they are for real."

Finally, find out whether the journalist interviewing you is a "mojo," short for mobile journalist. Also known as "backpack journalists," they will not only interview you for a print article, but also use a camera and/or digital recorder to produce video and/or audio versions. A mojo may work for a newspaper, television station, or Web news site. Prepare for a mojo by following the guidelines in the following sections on print, radio, and TV interviews.

Formulate Your Messages

A concise summary of your work and its importance should already exist in the form of a news release, prepared statement, and/or Web page. Use that material to develop your "elevator speech," says University of California San Diego PIO Kim McDonald. "Pretend you are in an elevator, and you are going down twenty stories, and you only have thirty seconds or a minute to describe to someone who knows nothing about science what you have done and why it is important," he says. "And the 'why it is important' is the most important thing, because people aren't so much interested in the details, because they don't really understand the science."

Besides developing a compelling elevator speech, also list the main points you want to get across in the interview. However, do not think of these discussion points as a script that you must religiously recite as you talk. Rather just keep them in mind during the interview, to make sure you include them. Among the issues to cover:

- What key concepts should people understand?
- Why should people care about your findings, beyond the brief statement in the elevator speech?
- What are your "Big Thoughts?" As discussed in more detail later, these include caveats, limitations, trends, controversies, and challenges regarding your research
- Are there actions an audience should take as a result of your work?
- How do political or policy issues affect what you can say? For example says veteran government PIO Leah Young, if you are a government researcher,

> Understand the messages of the day. The administration has a position, and you don't want to get into a situation where a reporter interprets what you are saying as the opposite of what the administration is saying. For example, if you find yourself being asked about the funding for your particular area, you just clearly say "There is only so much money to go around, and this administration has decided to put more of it into X or Y, and the administration has made the decisions that this is where they want to put their money."

Develop Artful Quotes

Developing quotes beforehand might seem calculating, since quotes are supposedly extemporaneous responses to a reporter's questions. But figuring out what to say and how to phrase it will protect you from stumbling over your words and saying something you do not mean. What's more, if you screw up a quote or do not correct yourself a reporter is perfectly within his or her rights to include that screwed-up quote.

So, figure out what you want to say about your work, perhaps jotting down key phrases that work well. In the process, try to tighten the phrasing of quotes. Such tightening can make a big difference in the impact of those quotes. For example, imagine if Franklin Roosevelt had padded his famous quote by saying "I think most of us would agree that these days the only thing we have to fear is fear itself, even with all our problems."

Prepare Anecdotes and Analogies

Colorful anecdotes enliven your research story and make it memorable to readers or viewers. So, think of anecdotes that tell memorable stories of your work's progress and discoveries. Editor Julie Miller cites an excellent example of an anecdote about research on African mole rats that made an article really stand out. "We led with the incident about this man who was searching for them and couldn't find them," she recalls. "He was sleeping in his sleeping bag on the ground and woke up with a crick in his back, because some mole rat had come up and made a mound right under his back in the middle of the night."

Similarly, think up concrete analogies to illuminate abstract concepts. The analogy of the changing pitch of the whistle of a passing train has long helped explain the Doppler shift, as has the venerable spiral staircase analogy to describe DNA's structure. Also, you may recall the sample metaphors in chapter 10—artificial dog, cosmic blowtorch, anaconda receptor, shotgun synapse, and jellyfish cells.

Quantify Vividly

While solid numbers lend your research story credibility, vivid numbers make your story accessible and memorable. Compare the size, speed, temperature, or other aspect of your research subject to something people can identify with. For example, as discussed in chapter 10, you could compare a tiny object to the diameter of a human hair (200 micrometers); compare a long distance with the earth's circumference, or compare a nanoliter-sized volume with the volume of a snippet of hair as long as the hair's width.

Stephen Maran, press officer for the American Astronomical Society, cites a memorable example of such vivid quantifying, which he dubs "factoids":

> The solar physics group at the Naval Research Lab was studying a particular class of coronal mass ejection that was more powerful than your ordinary coronal mass ejection. They said it's a billion tons moving at a million miles an hour. To me that is a weak factoid because nobody is familiar with a billion tons. So, Dr. Spiros Antiochos came up with the factoid that the energy contained in this moving mass hurled out of the sun was so many tens of thousands Nimitz-class destroyers moving at so many knots.

Failing to vividly quantify can deflate a good story, warns *New York Times* science writer Sandra Blakeslee. She recalls the unfortunate case of the Canadian ecologist who condemned the cutting of vast swaths of Canada's forests to make paper for American mail-order catalogs. "I said, 'You have such a good story to tell, since people could just as easily get all their catalog information online.' I asked him what is the mass of forest? But he hadn't done the calculations. He couldn't say anything like 'We are cutting down one Manhattan every two hours to feed your catalog habit.' So he had a beautiful story, but he didn't know how to tell it."

Think Big Thoughts

The more you can explain to a reporter how your research fits into the broader picture, the less likely that reporter is to misconstrue that context. So, before an interview, think about these ideas:

- How does your work address an important problem in your field?
- What are its caveats or limitations?
- How does it fit into the trends in your field?
- What controversies surround your work?
- What challenges lie ahead?

Such Big Thoughts can be more important to the reporter than the facts of your research, says science communicator Rick Borchelt. "The reporter is not calling you just to hear you spout your paper over again," he says. "The reporter is calling you for context and trying to figure out what that paper means—how to fit its information into what he has already put together."

Being candid about caveats and limitations will not reduce the chance of a story on your work. In fact, it will make your work more interesting to a reporter, says journalist/author Keay Davidson. "When somebody comes to me and says 'This might lead to a cure for cancer,' or 'This might lead to a solar cell that will generate ten times as much electricity as any other solar cell,' I immediately go on the defensive," he says. "If, by contrast, they acknowledge the problems and the limitations, that is very interesting in itself."

Also, being your own harshest critic makes you a more reliable source, says *Science News* editor Tom Siegfried. "It is a much more credible source who can point out reasons why what they found should be interpreted with caution and where the weaknesses in their study are. If they show that they know about those weaknesses, then you can trust what they say about everything else, so much more than if they hype their stuff and knock down other people's stuff, and don't apply the same critical standards to their own work."

Besides explaining how your work fits into your field, be generous in providing the reporter information on that field. Offer news releases, feature articles, Web sites, and other material that illustrates trends in your field. Such information might prompt the reporter to go beyond a news story to produce a longer feature article that will benefit you as well as your field.

Reveal Your Friends and *Foes*

Be prepared to provide reporters the names of both supporters and critics to comment on your work. You will not be able to shove your critics under the rug (or wherever you want to shove them), says Blakeslee. "I want them to be completely honest, because I guarantee I will find out," she says. "You may not want to mention your archenemy, but the reporter will get to that archenemy pretty damned fast. So, just be honest and say 'So-and-so has a different viewpoint, and you could talk to them for that viewpoint.' I know they're not trying to hide anything, and it makes me more sympathetic to their side."

Your own paper will likely contain clues to the contrary perspective that reporters seek, points out National Public Radio science correspondent Joe Palca. "I look for the sentence in the paper that says 'We used to think . . . ,' and then they cite the reference. And I call that reference and ask them 'You are wrong now. What do you think about that?' Also, I often will say to scientists 'Are you goring any sacred oxen with this work? Is there somebody whose career is built on saying the opposite?'"

In the interview, you can certainly go beyond a mere referral, also critically discussing the opposing viewpoint, says journalist/author Paul Raeburn. "You can say 'There is another group who has looked at some of this data and they've come to the opposite conclusion,'" he says. "You can explain what you think the problems are with the other group's conclusions. It helps us understand not only what you know, but how sure are you about what you know?"

Question Yourself

Once you have developed your messages, quotes, and other material, develop a list of questions you might be asked and practice the answers. Include the worst possible questions you might be asked. Your PIO can help you. Catherine Foster, formerly Argonne National Laboratories media relations manager, advises practicing "circular answers" to these questions. "If somebody asks a question, and you just answer it, then it is not a circle," she says. "But if you think in messages, then you answer the question and transition from the answer to your message. And you have brought the discussion back around again to where you want it to be."

Also, practice deflecting questions, says PIO Alisa Machalek of the National Institute of General Medical Sciences, who trains program officers in giving interviews. "The goal of a media interview doesn't have to be answering the questions of the interviewer, if they are inappropriate or out of the person's area of expertise," she says. "Many of these scientists are teachers and professors, and information exchange is very important to them. So, they are asked a question, and they answer the question. But they should also understand how to bridge from an interviewer's question—if it is veering into an area that they are uncomfortable with—to an area that they are comfortable with, and area that they can respond to."

Duke's Deborah Hill says she makes sure such practice sessions are rougher than the interview. "I will stand much closer than a reporter will, because it makes them uncomfortable; it offends their sense of distance. And I will put the lights right on them so it is hot. I make it much worse when they do it with me than it is with the reporter, so when they get to the interview it is a piece of cake. And they are well prepared."

Chapter 24 offers more detailed techniques for handling interview questions.

Translate Basic Research

You face a greater challenge in explaining your work to journalists if it is more basic. However, you can still communicate your work in a way that resonates with journalists and the public. First of all, says Blakeslee, point out that your work seeks fundamental insights that will ultimately contribute to applications. "It could be a story on the basic biology of protein folding," she says. "It seems to be kind of abstruse, but it is not. Protein folding abnormalities underlie Alzheimer's, Parkinson's, cataracts—many different diseases. So, relating protein folding to those disorders brings the discussion to why people should care." Of course, you should not overreach by implying that your basic findings may yield new treatments or new products if they will not—the dubious practice of what author Daniel Greenberg calls "may journalism," as discussed in chapter 10 on crafting news releases.

Besides citing possible applications, you can also legitimately present your basic work as just plain interesting for its own sake, says Maran. In fact, the details of a particularly arcane discovery might not matter as much as your general message, he says, telling astronomers "Unless you are dealing with serious science writers, your true objective should be—because it is the only thing that will ever work—to get across the idea that 'Astronomers like me are doing exciting and interesting things.'" He says, "If you seriously think that the story of your specific scientific breakthrough is going to matter to the reading—or especially the listening or viewing—public, you are just mistaken."

Be prepared for the loaded question basic researchers sometimes get from journalists: "Why should the public support your work when it has no immediate practical application?" To that question, an excellent rejoinder is "Astronomy (or particle physics, or neurobiology, etc.) inherently fascinates people, because it reveals the wonderful and exotic phenomena of the universe. Isn't it proof of that public fascination that *you're sitting here interviewing me for this story?*"

Finally, you also can assert that stories on exciting basic research have a value in attracting students to becoming scientists or engineers.

Do Not Triage Media

If you receive a lot of media calls—for example, when you publish your inevitable paradigm-shattering paper—you might be tempted to "triage" them. That is, you might want to relegate local reporters' interview requests to the bottom of your call-back list, in favor of those with major media outlets. *Don't.*

Rather, offer full access and cooperation to all media. One reason is that since the Internet is global, there are no more "local" media. The story in your hometown gazette will be listed by such news aggregators as Google News and Yahoo! News right along with *New York Times* and *Washington Post* stories. And, in fact, that local story might well be more comprehensive and explain your work better than the one in a national outlet. In any case, the more stories on your work, the more visible it is.

Also, local reporters will not always be local reporters, points out science communicator Lynne Friedmann: "Journalists are going to start off at the local publications, but as their careers progress, they go to the larger big-name publications. And when they get to those publications, they will remember you as somebody they know they can call."

You might also consider favoring mass media such as the *New York Times* or the *CBS Evening News*, believing they have more impact on opinion leaders than such professional magazines as *Science* or *Nature*. But those latter communications actually carry much more weight if your target audience is science policy leaders, indicated a study by Rick Borchelt and colleague Jon Miller. They identified 8,000 decision leaders in science and surveyed a sample to discover what information sources influenced them the most. The survey revealed that articles in *Science* and *Nature* and reports from national laboratories and scientific organizations carried far more weight than mass media. They found that these decision leaders ranked the *New York Times* in the middle of the pack and the *CBS Evening News* dead last.

Prepare the Reporter for Your Interview

To make your life and the reporter's easier, offer him or her background material beforehand. Do not expect to educate a reporter in the interview itself. The brain is not a bucket into which information can be poured, and if you deluge an unprepared reporter with background information during the interview, you will likely get an error-riddled story.

Thus, when the reporter calls to arrange an interview, offer to e-mail your paper, the URLs of pertinent Web pages, and other useful references. As discussed in chapter 7, your Web site should feature lay-level research summaries, visuals, past articles, FAQs, and other content that will prepare the reporter for the interview. Your FAQ can even include the questions you developed to prepare for the interview. It might pay off in that the reporter will ask those very questions, aiming the interview in the direction you want.

Inform the reporter about previous coverage of your work. This serves two purposes: it gives the reporter useful background and avoids surprises, as when an editor tells the reporter somebody else had written a similar article.

When a Reporter Calls Out of the Blue

When a reporter calls unexpectedly, do not feel you have to jump into an interview unprepared. Especially if the topic is complex and/or delicate, do the following:

- **Ask the reporter's name and affiliation, what story they're working on, and their deadline.**
- **Ask for any further information you need, such as a copy of a paper on which you are being asked to comment.**
- **Promise to call the reporter back well before their deadline.**
- **Call your PIO if you think he or she can help.**
- **Do a quick preparation as outlined above, taking time to think through the topic and the messages you want to get across.**

If the reporter insists on talking then and there, Gibbons suggests that "you have always got some meeting you have to go to, or a patient you have to see, and use that to beg off the phone and call them back before their deadline, after you have had a chance to think."

When Dealing with Media You Would Rather Not

You might receive calls from newspapers, magazines or Web sites that you would just not like to see publish a story on your work. Certainly, you do not have to cooperate with them. However, before you refuse, consider two factors: whether the benefit of the story will outweigh the pitfalls, and the fact that the media outlet will do the story with or without your cooperation.

Instead of kissing off the reporter, you might consider doing an interview, but with special requirements to protect yourself. Yarbrough recalls just such an incident, when she worked for the Arthritis Foundation. "The head of medical and scientific affairs actually brokered a deal with the *National Enquirer* in which he required that he review the story and edit it before it went to press. I thought it was very smart because it was based on the reality that a lot of people with arthritis read the *National Enquirer*. And why fight the newspaper when we could actually get some good information out there?"

24

Make the Interview Work for You

Now that you have prepared the messages and content of your interview, this chapter covers principles and techniques to help you in the interview itself.

First, before the interview even begins, decide who will participate and whether and how to go off the record. By all means invite colleagues to sit in whom you think can contribute to the discussion. However, do not crowd the room. Do not ask people to sit in just for political purposes. Also, understand that the reporter will want to quote only one spokesperson for most hard news stories.

Reporters generally prefer not to have a public information officer (PIO) sit in on an interview, although institutional policy might require it. At universities, PIO attendance at interviews is an exception, but in corporate and government laboratories it is usually the rule. More than just acting as a watchdog, a good PIO can be of considerable help in an interview, argues former government PIO Leah Young. "The PIO might think of a point that the guy just hasn't thought of and can write a note and slip it in front of him, or call attention to information that might have been missed," she says. If your PIO does not attend, at least he or she should be easily reachable to offer any help the reporter needs.

Do not make the interview do double duty by inviting other writers to sit in. You might just lose the story altogether. Science writer Robert

Cooke recalls such an instance when he interviewed a physician developing an implantable insulin pump for diabetics. When he arrived, Cooke was informed that a campus magazine writer also would sit in. "I really didn't like it," says Cooke. "It is hard to have an interview with other people asking questions, and it steers you off course. So, after doing the interview I just left, and I never wrote the story."

Assume you are on the record unless you and the reporter agree otherwise. Try not to go off the record at any point, because your off-the-record comment might find its way into the story—whether through misunderstanding or the reporter's malfeasance. However, depending on the reporter, you can negotiate off-the-record terms that will serve both of you. For example, *Washington Post* environmental reporter Juliet Eilperin told a workshop at the 2008 AAAS meeting for scientists on working with media:

> You can always say to a journalist "Can we talk off the record for a while, and I'm just going to explain to you what's going on, and we'll come back and figure out what it is you exactly want to use?".... Sometimes if you do that, it makes people more relaxed.... I'm perfectly willing to cut people slack, particularly if that's not what you usually do. And once you get beyond thinking every word that's coming out of my mouth is going to appear in a story, that sometimes helps people relax a little bit and talk in more conversational terms.

If you must go off the record and are unsure of the reporter's policy, preface your remarks by saying "I would like to go off the record here," and wait for an explicit assent from the reporter. When you finish your off-the-record remarks, say "I am going back on the record."

Besides going off the record you might also go on "deep background." The distinction is that off the record means that the reporter may use the information but not attribute it to you, while deep background means that the reporter cannot use the information at all, except for his or her own understanding. However, reporters' definitions of these terms may differ. If you think you will be dealing with either situation, determine before the interview how the reporter defines those terms and how you can invoke off the record and deep background.

You do have something of a "big stick" to persuade the reporter to adhere to background rules, says *New York Times* science reporter Sandra Blakeslee: "If the reporter screws up, they never have to talk to that reporter again," she says. "The reporters know this, too, and the reporters know that if they screw it up they are never going to get that source again."

Check Out the Reporter

As the interview begins, feel free to ask questions to gauge whether the reporter did homework by reading the paper and/or background materials you provided. You can assume that science reporters with major media will have prepared. However, the general local reporter might not have. Perhaps the material was too technical, or there might not have been time. You might be sandwiched between coverage of the local school board and the amazing cat that found its way home from the next state.

If you find that the reporter is not up to speed on your field, prepare to give a tutorial on the basics. And those basics might be *very* basic, says Blakeslee. "Every biologist knows the concept that protein shape determines function, but a science reporter—especially a rookie science reporter—or nonscience reporter might not know," she says. "I am always grateful when a scientist will just go over some fundamental concepts, because then you won't get the story wrong." Most veteran reporters do appreciate such tutorials, agrees Ben Patrusky, executive director of the Council for the Advancement of Science Writing. "Begin with first principles," he says. "I don't think you can ever go wrong with that. I have never been insulted by somebody telling me something simple."

But if the reporter appears to be uneducable, says Patrusky, you have the perfect right to ask them to go back to the drawing board…or rather the references. "If you feel it is hopeless, I think you have the right to say 'I just don't think you are equipped to tell this story,'" he says. However, emphasizes Duke communications director David Jarmul, respect the dedication of even the least-experienced reporter. "Even the twenty-three-year-old local reporter who is coming in from the fire, who doesn't know a protein from a tuna fish sandwich, wants to do a good job. You can help them. Just thinking they are a blithering idiot, which might be partially true, isn't going to be very helpful," he says.

General Interviewing Strategies

Once the interview is underway, there are strategies you can use to make the interview both painless and productive for you and for the reporter. Some general strategies:

- **Ask about time constraints.** Is the reporter on deadline, and if so, when must the interview end? Also, what is your deadline for getting follow-up points to the reporter?

- **Consider the reporter neither foe nor friend.** The reporter is not an adversary. Both of you want an accurate story on your work. However, the reporter is not an ally. The reporter works for the readers and viewers. So particularly if you are involved in a controversy, take heed of the advice in the next chapter on avoiding communication traps. Be neither paranoid that the reporter is out to get you, nor too complacent. Cautious cooperation is your best approach.
- **Assume lay-level understanding.** Assume you are talking to an intelligent lay person, unless the reporter is from the scientific media, and likely has more knowledge of your field. So, as you explain your points, make sure the reporter is following you by asking whether he or she understands.
- **Work within the medium.** Even though your findings have taken years to develop, the newspaper will only devote a few hundred words to them, and the television news segment 90 seconds or less. Design your explanation to fit within these limits. Think of such media stories as a headline service to alert audiences to your work, and that they can find more detailed information on your Web site and in your news releases and feature articles.
- **Maybe take a running start.** If you are new at being interviewed and feel uncomfortable, Blakeslee advises a running start. "One way I make people comfortable if they are nervous about talking with a reporter is that I will say 'We are going to talk about your work, but first let's back up, and go back as far as you want.'"
- **Or, start with your headline.** If you are an experienced interviewee or the reporter is on deadline, start with the headline and summary of your work and what it means. Forget your standard scientific communication structure of starting with background, methods, and data.
- **Avoid the "DTs."** That is "too much *data*, too much *terminology*," says Don Gibbons, communications officer at the California Institute for Regenerative Medicine. Researchers, he says, "are so proud of their research that they want to explain all of the intricacies, but it doesn't work. They don't understand that 'morbidity' is not a known term; 'phenotype' is not a known term; and death is not an 'adverse event,' it is death, a tragedy."
- **Speak deliberately.** As excited as you may be about your findings, control your talking speed. A fast tongue can lead to mistakes, sometimes embarrassing ones. Talk at a speed slow enough that a note-taking reporter can keep up with, and such that your tongue does not outrace your brain.

- **Skip equations.** "When people write equations for me, I'm dead," says Cooke, echoing the sentiments of most reporters. "The reason I'm a writer is I don't do numbers." You can, however, sketch basic diagrams, especially if you label them and give them to the reporter to take away.
- **Avoid too many "couching" terms.** Says Duke research communicator Joanna Downer, "If you look at a scientific paper or talk to a scientist, they use 'might,' 'possible,' 'perhaps.' In a single sentence they will have eighteen words that mean 'this may not be true,'" she says. "As a scientist, I know that it is very important for scientists to maintain that uncertainty in what they say. But, when they are speaking to be quoted, they have to limit that. You have to be as strong and controversial as you can be in order to be a useful resource to a reporter. And I don't think that it is necessarily impossible to be as strong as you can be and to maintain your integrity as a scientist."
- **Repeat yourself. And repeat yourself.** Repeat critical points and/or caveats throughout your discussion, to remind the reporter—especially if you sense from the reporter's question that they are being missed.
- **Be straight with facts.** Reporters have a keen instinct for ferreting out a story you want to obscure. If you hide facts, you will trigger that instinct, says Cooke. "As soon as you start stonewalling, reporters will start climbing that wall. If you hide something reporters will say 'Now, that is where the story is! This is the thing I need to lead this story with, because it is controversy, and that sells to editors.'" Also, points out journalist/author Paul Raeburn, you will lose the reporter's trust. "You can try to fool us, and you can try to mislead us or not give us a real honest sense of where the field is. And you might get away with it, and you might not. But if you don't, you are going to pay a high price."
- **Offer perspective on risk.** If your work reveals an increased health risk, put that risk in context, advises journalist Cristine Russell, a senior fellow at Harvard's Belfer Center for Science and International Affairs: "Go beyond just what your study said and try to think of useful comparative information, so that people will know what risk they are facing and how they compare it to other risks," she says. "Most journalists will not have the time or ability to find the absolute risk." Also clarify how large the risk—for example of a cancer—was to begin with? How many people are affected?
- **Avoid hype.** "Reporters have a keen nose for hype," says Raeburn. They understand that scientists can tend to overstate when they get excited about their work, he says. "We want them to be excited, because that keeps them going.... But they need to be a little bit careful about that initial

enthusiasm, with the tendency to overstate sometimes the significance of their findings."

- **Support your assertions.** A corollary to avoiding hype is to support your statements with cold, hard research. Be prepared to cite specific references to papers and studies to back up what you are saying.
- **Distinguish fact from opinion.** You are perfectly entitled to express an opinion. But be clear when you make an assertion based on data and one that is your opinion. Says Raeburn, "It is easy for us to confuse that. It's one thing to say 'Our data shows that sixty-two percent of people get better on this medication.' It is a different thing to say 'We believe this will be the most important medication to treat this illness.'"
- **Explain controversies fairly.** Give all sides of a controversy, including the views of reputable people who disagree with you. And while you should explain why you believe some researchers' views are not valid, do not slander them or engage in personal attacks.
- **Anticipate how your work could be misconstrued.** Beyond explaining the caveats of your work, spell out how the reporter or readers might misinterpret it. A classic example of such misunderstanding was the case of the "lost city" in the Andes reported in a 1985 release from the University of Colorado. Reporters, including from the *New York Times*, took the phrase to mean that the ruin of Gran Pajaten had indeed been lost, when it had actually received much attention by explorers, tourists, and media. The *New York Times* had to publish a correction to its story. At *Science News*, a staffer found the city on a roadmap, and the magazine published a story that said essentially "Lost City in Peru—Never Mind."
- **Make conflict of interest clear.** Just as your news release clearly states conflicts of interest such as corporate ties, also make them clear in interviews. Reporters today are acutely attuned to conflicts of interest, says Raeburn. "Now it is safe to assume that everybody you talk to is involved with some kind of company or somehow has some plan to profit from the research they are doing." Failing to clearly state your involvement with a company, for example, will sow suspicion with the reporter, he says.
- **Meticulously give credit and document that you have.** Your news release gave credit to collaborators, and so should your interviews. You need not laboriously recite every contribution, but make clear what you did in your laboratory and what your collaborators did. In fact, such information is useful to the reporter. It tells the reporter that, for example, that while you are the person to ask about the laboratory analyses, your collaborator knows more about the animal studies.

- **Let the reporter put words in your mouth.** Since reporters communicate for a living, they might come up with good phrases or analogies, says *Science News* editor Tom Siegfried. "You have to try to put words in their mouths, both for purposes of your own understanding and for purposes of helping them articulate more clearly."
- **Check the reporter's understanding.** Periodically ask the reporter if he or she understands and whether you are saying what needs to be said for the story.

Strategy for Questions

You have much more control over questions during the interview than you might think. These tips will help you manage questions, and, in fact, also help the reporter get a better interview:

- **Answer questions you want to have been asked.** Feel free to volunteer answers to questions the reporter has not asked. Thus, you will provide information that will make the story better.
- **Learn to "block and bridge."** If a reporter's questions veer away from what you think needs to be said, "block" the undesired direction and "bridge" to the point you want to make. Gibbons offers these blocking and bridging phrases that he learned from colleagues at the Centers for Disease Control and Prevention:
    - "What I think you are asking is..."
    - "The real issue is..."
    - "Let me put that in perspective..."
    - "What's important to remember is..."
    - "I can't discuss...but I can tell you..."
    - "It is true that...but it's also true that..."
    - "Actually, that's not true, but what is true..."
    - "I don't know about that...but what I do know is..."
    - "What I really want to talk about is..."
    - "Your readers/viewers need to know..."
- **Avoid negative buzzwords.** A reporter might ask a negatively phrased question, but you can avoid being caught by it, says Johns Hopkins PIO Joann Rodgers. "So the journalist asks 'When did you stop beating your wife?' You don't have to answer that question. "You can say 'Hey I love my wife,' avoiding the buzzword."
- **Answer any legitimate question.** Even if you prefer not to talk about some aspects of your work, if the question is legitimate, answer it. As

discussed in the previous chapter, you should have prepared answers to those questions.

- **Say if you are not certain, and why.** You are not omniscient, and your work has uncertainties. It is perfectly all right to admit that. However, explain the reason for gaps in your knowledge or certainty.
- **Never just say "no comment."** A terse "no comment" is actually a comment. And as indicated in chapter 25 on avoiding communication pitfalls, it can get you into trouble. Try to explain why you cannot talk about something, so the reporter will not quote you as having no comment. For example, tell them when their question covers findings that are still not conclusive or are in press.
- **Prepare for the application question.** As discussed previously, your messages should include potential applications, even if you are a basic researcher. A good answer can avoid embarrassment. For example, there was the case of the Argonne National Laboratory researcher who performed DNA and lead content analysis of Beethoven's hair and skull fragments. In a possibly embarrassing question, one of the scientists was asked why the federally funded laboratory should concern itself with the long-dead composer. Recalls Catherine Foster, ANL's media relations manager at the time, "He explained how the techniques that they had developed to do this work were now being used for environmental studies. He went into some detail about the environmental work that he was doing and handled it beautifully."
- **Tolerate repetition.** A reporter might ask you the same question several times. Rather than being annoyed, see this as a sign that you are not explaining your work clearly or giving good quotes. Says Russell, "I am persistent. I am a nice dog yapping at their heels. I do keep trying to ask the question either until I can understand it, or I can get a decent quote that will help the reader/listener to understand."
- **Beware the "Danger, Will Robinson!" question.** Answering what seems a perfectly innocuous question might lead to an embarrassing gaffe. National Public Radio science correspondent Joe Palca recalls just such a close call when he was working for a scientist studying the parallels between sleep and hibernation. A *Discover* magazine writer asked whether the research meant that humans could hibernate. "It stopped me in my tracks and I realized *Danger, Will Robinson*! that I was asked a perfectly reasonable question but the answer if I didn't say it right would make me and the whole research project look imbecilic." So, if a question pushes

your alarm button, take some time to consider how an answer might be twisted into an embarrassing headline.

- **Do not fall for the silent treatment.** Reporters sometimes use silence to get you to expand on an answer beyond what you wish to say. Rather than falling for this silent treatment, ask the reporter "Do you have any other questions?"
- **Know when to shut up.** Discipline yourself to stick to your messages and do not blather, says Gibbons. "You made your three points, you have a meeting to go to, get off the phone." You might even ask for help in shutting up. When Gibbons suspected one administrator might not realize when he said enough, he had the administrator's secretary buzz him after an appropriate time, so he could excuse himself gracefully.
- **Expect fact checker's questions.** For magazine articles, you could get a call from a fact checker. You might have to explain aspects of your work again, since the fact checker is coming into the story cold. Be gracious, but do not tolerate total ignorance, says Sally Maran, longtime editor for *Smithsonian* magazine. "The scientist should not have to suffer fools, and my researchers at *Smithsonian* were always told that they need to do some background reading, so they can ask intelligent questions," she says. "And if they get a fact checker who is obviously biased in some way or is wasting their time, they would not be out of line to call the editor and ask, 'What is up with this?'" Editor Julie Miller warns that the reporter might also call with eccentric follow-up questions from his or her editor. "There is just no telling what crazy questions editors are going to come up with. If they don't have the answer in their notes, the writer is going to have to call back and say 'I can't believe it, but my editor wanted to know what color shirt were you wearing that day?' You have to be there to answer, or if you are not available, to empower one of your grad students or a colleague."

Strategy for Additional Information

Besides providing your messages and background materials, you might offer the reporter

- **Links to visuals and multimedia.** Because reporters are under more pressure to gather visuals, they appreciate links to Web sites and information on sources of diagrams, animations, and videos. Even though the artwork might not be directly useful, it guides the art director on designing illustrations for the story. And the newspaper or magazine Web site might just link directly to your Web sites and videos.

- **What things look like, feel like.** Such sensory detail can enliven a story and make it more interesting to readers and viewers. And, it can make the story more interesting to editors, who control where a story appears in a publication or newscast. For example, let the reporter see/feel a sample of your research material—unless, of course, it is toxic.
- **Tours and demos.** Remember that "News is what happens to reporters." Make more happen to the reporter than just an office interview, such as a tour or demo that could make for a livelier, more interesting story. Such experiences might not be useful for a print reporter on deadline, but they are all but required for a television news segment.

Strategy for Style

While your research provides the substance of your story, your interview attitude and style will help the reporter tell your story interestingly. Some tips:

- **You are the good guy.** Going into an interview, you have an advantage over, say, a businessman, lawyer, or politician. As a researcher you have an inherent credibility. Remember the Harris poll discussed in chapter 1, which showed that doctors, teachers, scientists, and professors are highly trusted professions. So, unless the interview covers controversy or malfeasance, expect it to be a fairly benign, even enjoyable event.
- **Get personal.** The cool, impersonal image of researchers is a barrier to communication that you can overcome by revealing yourself as a person, says Jarmul.

  > Readers are interested in people and scientists are people. And when scientists insist on being seen as somewhat faceless intellectual pursuers of truth, that's fine, but it is not terribly interesting to read about. I have always thought it is a better story—and it is better for science—when we see scientists as living, breathing, imperfect people, which we know them to be.... Scientists understandably are concerned about being trivialized and losing control or some of their dignity, but they're actually more effective when they let down their guard a little and share their humanity with us.

- **Push your enthusiasm button.** Feel free to show your excitement about your work. Sandra Blakeslee quotes her late father, the pioneering science writer Alton Blakeslee: "My dad always said 'Push your enthusiasm button before you begin to write a story,' and I think the researcher should push the enthusiasm button, too."

- **Use fun words.** You might be tempted to be more serious and formal in an interview than, say, in the lunchroom with your colleagues. However, colloquialisms can help the interview, says Blakeslee. "What I love is when the scientist uses these fun words. They will say 'that is chockablock full,' and the word 'chockablock' is not used very often, but it makes a perfect quote. They don't want to sound silly and have things taken out of context, but the use of lively colloquialisms can make a science story better and it can make them more human."
- **Use humor.** A little humor can memorably get across a point about your research, just as the cartoons in this book memorably make points about communication. For example, Miller recalls how humor helped an esoteric story about animal hormones for *Science News*. "They were talking about an enzyme that worked two different ways, and the researcher said, quoting an ad, 'It was like *It's a breath mint! It's a candy*!' The writer went 'Great! Here's something I can use that gets the idea across that is a little bit humorous.'" However, when you use humor, expect it to be quoted. Unfortunately, says Miller, the scientist was chagrined when he saw his words in print "because he thought it made him look stupid. He denounced our writer at a scientific meeting because of that."

Follow-up Strategy

Taking these following simple follow-up steps can make a big difference in the accuracy and success of the story:

- **Give the reporter a pop quiz.** "It is perfectly permissible to ask a reporter to feed back what you just said," says Rodgers. "As a journalist I used that all the time when I was doing hard science. I would get through the whole interview and I would say to the scientist, 'I am going to tell you what I think you said; what the story is.' There were many times that I got it wrong, that I had gotten distracted, waylaid, I didn't get the main point; and the scientists to a man or woman were so grateful that I had done that so they could correct it."
- **Be available for follow-up questions.** Give reporters your e-mail address, cell phone number, and any other contact information to enable them to

ask follow-up questions. Also, as discussed previously, designate alternate spokespeople if you are unreachable.

- **Offer to check facts or quotes.** Reporters may, or may not, be willing to check facts or quotes, but it is worth offering to do so. Policies vary among publications. Some professional magazines such as *Science* may allow a researcher to read a draft for accuracy. Newspapers almost never share drafts. Also, individual reporters may have different policies. For example, says Eilperin, "I'm willing at that instant before you hang up the phone to go through people's quotes, and I often say 'Look, I'm not going to pick which of these quotes I'm going to use, but I'm happy to say what are those I'm considering.'" In asking for a review, understand that you do not really have the expertise to judge journalistic articles, says Russell. She warns researchers "You are biting off more than you can chew.... Think how you would feel if I said, 'I think I should write your study up for the journal. I think I would do a better job with that.'"
- **Follow up with an e-mail.** If any points might have been missed or misconstrued, send the reporter an e-mail of thanks for the interview, and include that you wish to clarify points that you might not have explained clearly. Especially if the reporter is not a science reporter, you might e-mail the messages and caveats you prepared for the interview. Send a copy to your PIO and colleagues.
- **Send a congratulatory note.** Once the story is published or broadcast, if it was well done, send an e-mail commending the reporter, says Blakeslee. "That will do two things. First, it will open the door so the reporter will come back to you for being a source on other stories. Second, it just sets up a relationship." Such a friendly relationship, says Blakeslee, can last for many years, giving the reporter an invaluable sounding board on stories in your field.
- **Expect errors.** Media stories on your work will contain errors. Live with them, says Jarmul. "If you end up with a story that is ninety percent right, accept it for what it is and deal with it and move on, unless an error is truly egregious and harmful. In the totality of the situation, it is probably going to be just fine."
- **Correct serious errors.** If an error is serious enough to compromise the story's accuracy or your reputation, ask for a correction. "I don't think scientists should be reluctant, feeling that it is beneath them to complain," says Russell. "If it is something that is egregious, their editor should know, and it then won't happen again." Russell points out that such complaints can serve a larger purpose by not allowing incompetent reporters to get by with poor work. Start by simply asking for a correction, but if the error

is serious and you do not receive satisfaction, graduate to a letter to the editor.

- **Keep the door open for further stories.** "Reward the good reporters," advises Russell. "Why not reach out to a single reporter and say 'I have some interesting stuff going,' and encourage that reporter to do an exclusive or enterprise or project piece with them where they can have a little more time and where they can get into the process of science?" Or, says Russell, just issue an open invitation for the reporter to come back and visit your laboratory. Of course, as indicated previously, do not offer exclusives on hard news stories.

Giving Radio and Television Interviews

You may be less comfortable doing radio or television interviews than giving print interviews. They require more showmanship and convey less information than does print. However, radio and television segments can still get across the basics of your research, and with only modest preparation you can do a perfectly fine job.

Study and Rehearse

To prepare for your television interview:

- **Familiarize yourself with the medium.** Critically listen to radio or watch television news and interviews to observe how interviewees perform. Study how they use gestures, inflection, and examples to get across their points. Mimic the habits of effective people and avoid the habits of ineffective ones.
- **Prepare your statement.** Script yourself by developing radio/TV versions of your messages, quotes, and analogies. Limit the length of messages compared to the print versions, because you will have less time.
- **Consider the listeners.** Ruthlessly rid your script of jargon, which can be especially toxic to a radio or television interview, warns Tinsley Davis of the National Association of Science Writers. "With print, it is easy to take the jargon out," she says. "An editor will scribble it out. But when you are doing TV, it kind of flows out of your mouth, and you tend to go back to the way you are used to explaining things."
- **Practice.** Record mock radio or television interviews, in which you answer prepared questions and deliver your messages. Your PIO or media

services department can help you. Ask colleagues, friends, and family to critique your performance—emphasizing that you want constructive criticism, not just applause.

- **Prepare a "media bio."** Produce a succinct bio statement that the interviewer/host can use to introduce you. Make it sound like an introduction; for example, it is fine to use subjective terms such as "leading authority" or "prominent scientist." Of course, be truthful.

Effective Video Interviews

To make your television interview effective:

- **Watch the show in advance.** Pay attention to the host's interviewing style, segment length (so you can know when to offer final points), use of props, and color of the set (so you can dress accordingly).
- **Bring your own questions.** Give the producer a list of the questions you think will make the best interview for both you and the program.
- **Show the cow.** Plan visuals such as backdrops and props with the producer and figure out how to use them most effectively. Remember the adage about video: "say cow, see cow."
- **Consider makeup.** A buff of powder may be all you need to reduce face shine. However, given the resolution of high-definition television, make sure it is not too heavy. For studio interviews, rely on a professional.
- **Dial up the energy of your voice, expression, and gestures.** Most researchers come across as flat on television, because they are not used to "projecting through the screen."
- **Be concise in your answers.** Allow for the fact that a television news segment will not run more than 90 seconds, and that the average sound bite will be about nine seconds.
- **Control the interview.** Remember that you can stop and start over in taped interviews. So, if you fluff a line, pause, then announce that you are trying again, and start over.
- **Do not be funny or subtle.** Television interviews must be straightforward.
- **Use the host's name in the conversation.** It creates a sense of intimacy and involves the audience.
- **Be still.** Do not fidget, fiddle with your hair, touch your face, wiggle your foot, or chew gum. Do not swivel in a swivel chair, or sway when standing.
- **Smile.** Look pleasant, even if you are bringing bad news. Smiling portrays you as sympathetic and increases viewer interest. For example, Mellody Hobson, financial contributor for *Good Morning America*, is always smiling

brightly when she comes on camera. Her demeanor makes you look forward to a pleasant interview, even when the economic news is not.

- **Watch your eyes.** Do not look at the camera for taped interviews, but at the reporter. Sometimes in remote interviews only the camera is present, with the reporter asking questions over the phone. If so, look at a person seated beside the camera. An exception: look into the camera if you are being interviewed live from the studio. When in doubt, ask the director where to look.
- **Use good posture.** When seated, sit up straight and lean forward slightly in the chair. This position makes you appear more forthcoming and dynamic. Sit on the back of your suit jacket to pull down shoulders and eliminate a fabric hump. When standing, keep hands at sides, not in your pockets or with arms folded. Button your coat when standing; unbutton it when sitting.
- **Do not wear sunglasses or tinted glasses.** They make you look slightly sinister. Also, eyes are very expressive, and you will lose that expressivity with glasses.
- **Dress for television.** For most news interviews, your everyday clothes are fine, as long as they are neat and conservative. Avoid shirts or jackets with small stripes, checks, or herringbone pattern. They shimmer on video, causing a distracting moiré pattern. For more formal occasions, men should avoid very dark suits, particularly in combination with white shirts which can make your face look pale. Gray suits and pastel shirts are more flattering. Wear over-the-calf socks so your leg will not show when seated. For women, dress simply in neutral colors, with no boldly patterned scarves or blouses, no large or clanky jewelry, and no short skirts.
- **Freeze at the end.** In a studio, stay in place after the interview. Credits may be rolling over the scene. Wait until the director tells you that the interview is done.

Effective Radio Interviews

Some tips on giving effective radio studio or phone interviews:

- **Provide quality sound.** For phone interviews, use a handset and landline, not a cell phone or headset. Turn off call waiting, so you will not get beeps during the interview. Try to use an ISDN line. Such a digital line, which your news office or radio/TV office might provide, transmits broadcast-quality sound.

- **Listen to radio programs and podcasts.** As with television programs, listen to the program or podcast on which you will be featured, so you can match its style and content. Also, note the length, so you will know when to make final points. Think about the types of stories being featured. For example, Steve Mirsky, who produces *Scientific American*'s podcasts, seeks "to have an entertaining, informative interview for the audience of science-interested people. I want to know what other reporters have not bothered to ask you that is worth talking about." Knowing such objectives can clue you on how to approach your subject. Besides *Scientific American*'s podcast, the online reference section lists radio programs and podcasts from the American Chemical Society, IEEE, *Nature*, NPR, *Science*, and *Technology Review*.
- **Bring your own questions.** As with television shows, providing questions can help both yourself and the program.
- **Always assume you are being recorded.** Audio recording is less obvious than video recording, so you might not realize that everything you say is being recorded. So, whether giving a radio interview in your office, over the phone, or in a studio, assume you are being recorded.
- **Do not fear the microphone.** A live microphone tends to intimidate people unaccustomed to giving interviews, says Davis. She recalls one unfortunate case of a researcher who reverted to technical jargon before the microphone. "As soon the microphone was turned on, it was like she was back in the lab...the microphone just made her go back into what she was comfortable with; what came out of her mouth in all those seminars she had given." A bit of practice with an audio recorder will help you overcome microphonophobia.
- **Record yourself.** As with television, record yourself practicing your radio interview, so you can correct any verbal tics or glitches. Also, record your actual interviews using a phone tap, so you can critique your performances and also post them on your Web site.
- **Watch your diction, accent.** Ask your PIO to tell you frankly whether you speak clearly enough for radio. If not, articulate your words more carefully and try to correct your accent. "We can't do subtitles," says NPR's Palca, although he notes that radio correspondents do have techniques to get around a thick accent. For example, one accented researcher declared that a drug he was studying relieved "acute *boots of pan*," recalls Palca. To clarify the accented words, Palca repeated them in his question, "When people experience these acute *bouts of pain*..." Such tricks aside, if your accent might trigger "boots of pan" in listeners, consider asking a colleague who speaks more clearly to do radio interviews.

- **Monitor your sound level, quality.** Once the interviewer establishes your sound level, try to control your voice to maintain that level. Also, keep a constant distance from the microphone so the level does not fluctuate. For radio and television interviews, make sure that clothing does not cause rustling over a lavalier microphone.
- **Quiet your environment.** For phone interviews, clear your desk so you will not bump anything. Close the door and turn off cell phones, pagers, and computer sound. Warn people not to knock by putting a note on your door or telling your assistant.
- **But also "sound-enrich" the segment.** Radio reporters do like ambient sound when it adds to the sense of immediacy. For example, Palca once interviewed NIH director Harold Varmus peddling his bicycle through traffic in Washington, DC. The ambient sound helped portray Varmus as an iconoclast who did not care about the niceties of Washington bureaucracy.
- **Provide sound effects.** You are absolute gold to a radio reporter if you happen to study sound-producing topics like bird calls or explosions, and provide those sounds, says Mirsky. "If you come equipped with sound, you are a hundred times more likely to get my attention," he says. Even if you do not have compelling sound effects, providing ambient sounds of your laboratory help give radio segments a sense of immediacy. Your radio technician can help record those sounds.
- **Help the reporter help you.** With an experienced radio reporter, feel free to concentrate on explaining your work compellingly. The reporter can edit around any stumbles or lapses into argot. For example, says Palca, "I will find the twenty seconds of nonjargon stuff in the middle of your jargon-laden interview." In fact, says Palca, your sound bite may be no more than a 'bridge' that enables the reporter to add narrated explanation. Says Palca, "Their sentence may turn out to be '…and we thought we were dead in the water at this point, and then we had this great idea.' That is the cut; then I tell listeners what the scientist said."
- **Keep it energetic.** Remember that on the radio your voice is the only way to project your enthusiasm and personality. Vary your tone, inflection, and pace to maintain interest. Standing up during a phone interview is one way to increase energy. Standing increases alertness and makes your voice stronger and your inflection more dynamic.
- **Describe visual things.** Craft vivid descriptions of important visual aspects of your work. For example, it is more interesting to declare "We use this giant, shimmering research balloon that looks like a jellyfish as big as a house when it takes off," than to simply say "We use a large research balloon to carry our instrumentation."

- **Use vivid verbs.** It's better to say "I tore down the road like crazy," rather than "I made my way down the road."
- **Use short, declarative sentences.** Rambling loses listeners.
- **Use the host's name in the conversation.** As with television interviews, it creates a sense of intimacy and involves the audience.
- **Avoid being ambiguous or subtle.** It will likely be lost on radio audiences, who tend to be casual listeners.
- **Keep a glass of water or a hard candy nearby.** These will be helpful should you get a cough. If you must cough, try to mute the microphone or telephone handset, or at least do so discreetly.
- **Use a cheat sheet.** With radio, you can have a cheat sheet in front of you that includes your main points, the station call numbers, the host's name, city, state, and the station phone number in case you are disconnected. For call-in shows, write down the names of callers so you can refer to them by name.

Doing E-mail Interviews

Reporters use e-mail interviews when they cannot easily reach scientists by phone or in person, and/or when they only need a brief comment. For e-mail interviews, use the same basic guidelines as with phone or in-person interviews: get your messages across, and supply the analogies and other content that help a reporter do a good story. And write to be quoted, emphasizes Blakeslee. "People's typed answers are more stilted, usually not as spontaneous and extemporaneous," she says. Blakeslee advises writing colloquially, as if you were talking to a friend. To test whether your answers are conversational read them aloud.

You might also send your e-mail reply to colleagues and/or your PIO before sending it to the reporter, to make sure your answers are both accurate and engaging. However, do any checking quickly, so you can transmit it before the reporter's deadline. Also, assume your e-mail will be read by the world, even though you might specify that certain parts are to be off the record. Experienced science journalists will no doubt hold your replies in confidence, but there may be reporters who do not.

Working with a Narrative Journalist

Besides news and feature articles, some journalists also do long narrative pieces of the type published by the *New Yorker*. For such a piece, the journalist will want to spend many hours in your laboratory and to interview you and your colleagues at length. The resulting article will be novelistic, in that it will explore not only the concepts of your research, but the personalities, tribulations, and triumphs surrounding it.

While a narrative article will take much more of your time and effort, the result is well worth it, says journalist/author Jon Franklin, two-time Pulitzer Prize winner. "Readers love it, and they read it, and they remember it; and the gestalt they get is that science is a human process," he says of the narrative article. He points out that a narrative article can produce a much greater impact than a news or feature story. "A lot of newspaper editors and readers tend to think if you get something in a paper then it's been read," he says. "What they don't realize is that most people, if they read five percent of the paper you are lucky... So ninety-five percent doesn't get read by anybody except maybe the writer's mother and the subject." In contrast, says Franklin, "A narrative is an experience. The way we prefer to learn—the best way to learn—is by experience.... Experience changes you. You achieve suspension of disbelief. The person lives in your story, and when [the reader is] done, no, it is not as indelible as if it had been a real life experience, but it is as close as you can get."

Franklin cites as an example of the impact of narrative journalism, a narrative story he did on a public health nurse who worked with tuberculosis patients. "My managing editor... came back and said, well, she liked that story; at least it broke some news. She said 'I hadn't heard about tuberculosis being so scary.'" However, when Franklin pulled the past clippings of news articles on tuberculosis, the result was a thick file. "The managing editor, who was always on us to read the paper, didn't know it had been in the paper," says Franklin.

To achieve such impact, the narrative journalist will want to eavesdrop on the life of your laboratory, often over weeks or months, says Franklin. "When I go in as a narrative writer, the first thing I do, talk to everybody about what they are doing, and why they are doing it, and what they think is going to happen," he says. "I want to watch something happen I don't want to be told about what happened last year."

Unlike the news reporter, the narrative journalist might seem somewhat aimless, because he or she likely does not know from the beginning what the story will be, says Franklin. So, when working with a narrative journalist, allow for that seeming lack of direction. Also, the narrative journalist will become more personally involved in the story—necessary to make the story compelling. For example, Franklin vividly recalls one story in which he covered a Duke neurologist and his work for many weeks. However, the story was not working for him as a coherent, compelling narrative. "I felt like I had lost my own best friend and thought at the time it was going to be a disaster," he recalls. "I was going to get fired; my children would hate me; the world would throw ripe tomatoes at me. Until finally it hit me: 'Oh, it is a love story!'" Franklin realized that telling the medical story as a narrative meant telling the story of a patient, Judith Vertucci, and her husband Jim. The resulting five-part series was published in the Raleigh *News & Observer* in 2001. The online resources for this chapter at *ExplainingResearch.com* include a link to a collection of Franklin's *News & Observer* articles, which constitute excellent exemplars of narrative journalism.

Says Franklin of his stories, "I am trying to get the life of the lab and the life of science. And I am really trying to get it the way it is—not the way I think it is, and not the way they want it to be. Not the way anybody expects. Very rarely in these stories do they come out like anybody expects."

Participating in News Conferences

News conferences can be an excellent way to explain your research to many journalists at once. They will all hear the same explanations, the same answers to the same questions, and will receive the same background information.

In deciding whether to call a news conference, the rule should be "When in doubt, don't." That is, you and your PIO should call a news conference only if you have breaking news, and only if so many journalists are interested that the most efficient way to serve them is to meet them all together. However, if only a few reporters are requesting interviews, it is still best to handle them individually.

There are different levels of news conference, according to the number of media you expect:

- A full-blown news conference for stories of great interest, held in a meeting room or hall, with formal questions and answers
- A smaller media briefing, convened around a conference table, for stories of more moderate interest
- A press availability for a story that might attract few journalists. This is an event in which you agree to be available at a certain time, should reporters wish to schedule an interview

Your PIO will manage the logistics of calling a news conference and contacting journalists, so you need not be involved in that process. However, if you do find yourself calling your own news conference, chapter 26 covers how to organize and run a news conference at a scientific meeting. The same basic guidelines apply for an individual news conference.

Prepare for a news conference much the same as you would for an interview. Develop messages, analogies, quotes, visuals, props, and even demonstrations. One difference from an interview, however, is that you should consider reading prepared opening remarks, in which you explain the topic of the news conference as you want it explained. Such remarks give you the chance to make your points up front, rather than having to rely on reporter's questions. Such opening remarks are especially important if the news conference involves a controversy. Write these opening remarks so that they are conversational and quotable, and practice them so that they sound natural and not recited. Give out copies of your opening remarks, along with a news release and background material, including a FAQ and your Web site URL.

Consider having a practice session, in which your PIO or others ask you questions. Foster would schedule just such sessions: "We would bring in other communicators and have them ask questions, and we would tell them 'One of you be mean, and one of you be stupid.' I would tell the researchers 'What we are going to do is going to be more painful than what you are going to get, but you will be more ready because of it.'"

In the news conference, ask the PIO to recognize questioners one by one after you have read your opening remarks. Most reporters are used to shouting for attention in news conferences, but especially when the topic is controversial, recognizing individual reporters makes for a calmer atmosphere and in fact, better information exchange. Reporters can be persuaded to follow such guidelines even for highly charged news conferences. For example, I asked reporters to be recognized to ask questions when I conducted a contentious news conference with Nobelist Richard Feynman at Caltech in 1986. Feynman wanted to explain his contrarian position as a member of the committee investigating the cause of the space shuttle *Challenger* disaster. I told reporters that asking them to raise their hands to ask questions would not only improve information exchange but also enable each question to be spoken into a microphone for the benefit of reporters listening in on a phone line. Also, I allowed each reporter a follow-up question. Reporters grumbled at first, but went along. During the conference, everybody got to ask their questions, and Feynman, who loathed giving media interviews, was much more comfortable with the process. In fact, I believe that his comfort led to better stories for the reporters, since he was more willing to provide information in the less boisterous atmosphere.

Also, limiting the news conference to one hour reduces stress. An hour seems to be a length sufficient to give reporters a chance to ask all the questions they need to, while being about the limit of energy and patience for many researchers.

If you are distinctly averse to news conferences, however, by all means ask your PIO to make other arrangements to provide media information. For example, when Caltech's Roger Sperry won the Nobel Prize in 1981, he refused to make himself available for a news conference or give any media interviews. He later explained to me that he suffered a nervous disorder that caused him to react extremely to any stimuli, such as social situations. So, one of his colleagues stood in for him at the Caltech news conference to explain the work that won him the prize. And he later agreed to answer written questions from reporters, conveyed through me.

25

Protect Yourself from Communication Traps

For all the benefits of communicating your research, there are also pitfalls. You might be misquoted or accused of hogging the credit or hiding corporate ties. Or you might find yourself swamped in the tidal wave of a hyperstory that feeds on itself ad infinitum or deciding whether to tiptoe through a delicate story. This chapter offers techniques to extricate yourself from such traps or, better still, to avoid them altogether.

Give yourself a head start in avoiding some communication pitfalls by understanding your institution's communication policies and culture—as recommended in chapter 8 on developing your communication strategy. For example, if you know the policies on portraying animal research, you can avoid the predicament of the neurobiologist described in chapter 8: ignorant of the university policy prohibiting cameras in the vivarium, he allowed a journalist's crew to photograph monkeys in restraining chairs. Also, as recommended in many other chapters, get to know your institution's communicators and how to work with them. The online section on working with public information officers (PIOs) at *ExplainingResearch.com* offers a good guide.

Beyond understanding the policies, also understand your responsibilities in carrying them out. For example, Johns Hopkins PIO Joann Rodgers bluntly instructs the university's epidemiologists about their role in communicating to the public, telling them:

> As part of your scientific training and your work in a science-based organization, you will be called upon to be a citizen of your institution. Your expertise is important for your field, your research, and your patients' care, but you also have to be the face of the institution when something goes bump in the night.
>
> You are going to be locked in small rooms for many hours with lawyers and communicators, crafting messages in very careful ways. You can't wing this. Just because you are smart in molecular biology is not going to automatically make you smart in how you say something. You have to work at this. You have to put in the same effort at honing a message that you do in honing an experiment. You won't like it, because it is going to take five times longer than you want. Such a crisis is only going to happen, we hope, maybe only once or twice in your lifetime, but you have to be prepared for that.

You can prepare for that "bump in the night" by brainstorming with your communicators how to handle the worst-case scenarios involving your research—whether death of human subjects or research animals, a radiation or biohazard leak, a political firestorm over a controversial finding, or a high-profile hyperstory. Work with them to formalize the communication plan, in particular what role you will be expected to play. Later sections in this chapter cover in more detail how to plan for hyperstories and manage crises.

Preemptive communications can help "immunize" your research against such problems. For example, science communicator Cathy Yarbrough relates how she immunized the Yerkes Primate Center against any claims that the center was secretive: "We wanted not to have the reputation of being a closed-door facility," she says. "I gave tours to community and educational groups, and I kept really good records, so I could say, for example, that I personally gave tours to a thousand people. And in talking to reporters, I could counter claims of animal rights people that we ran a closed facility by reminding them who from their media organizations had been here in the last year."

Another example of such immunization was the communication planning for the release of a genetically engineered virus into a cabbage field at Cornell in 1989. The virus, developed at the Boyce Thompson Institute for Plant Research, was designed to target the cabbage looper insect pest. As science editor at Cornell at the time, I worked on the team that created communications on the project. We decided to make the entire process transparent. We issued a comprehensive news release covering the experiment. We also prepared a press kit with a Q&A explaining what reporters would see when the virus was released. The Q&A aimed at heading off any misconceptions about the experiment. For example, we explained that the scientists wore isolation suits not because of any danger, but to avoid spreading the virus to an adjacent control field. Beyond media relations, the Boyce Thompson administrators met with local officials to inform the community of the experiment. The U.S. Environmental Protection Agency gave presentations to interested environmental and citizen's groups. And on the day of the release, we invited reporters to witness, photograph, and video the virus release.

As a result of all these efforts, any potential controversy was defused. As Anne Simon Moffat reported in *Science* magazine, "The experiment might have been a public relations nightmare since it was the first time a recombinant virus, albeit a disabled one, had been introduced into open fields. Yet there was minimal controversy about the spraying, and local and national media coverage was largely positive." Moffat quoted Boyce Thompson director Ralph Hardy as saying the goal of the communications "was to operate in an open manner and to make sure that the people in the local community and in a broader area were informed early and at each significant step." As Moffat concluded, "The strategy paid off."

Monitor Your Reputation

Monitoring what is being posted about you on the Web can help protect you from being blind-sided by a controversy. Some useful steps:

- **Use the Google Alerts service to monitor Web pages, blogs, video, and discussion groups and automatically notify you by e-mail of news stories and news releases about you, your center, your research topic, and so on.**
- **Use services such as Technorati and BlogPulse to subscribe to RSS feeds to keep track of blog posts mentioning you or your topic.**
- **Determine how influential a particular blogger or Web site is using Alexa to obtain statistics on traffic.**
- **Monitor what is being posted about you on Twitter by using the service Twilert, which sends regular e-mail updates of tweets containing designated key words.**
- **Subscribe to online groups on your topic to directly monitor and participate in discussions that involve you and your work.** Among the sites to check for relevant groups are AOL, Google, MSN, and Yahoo!

Of course, if you do discover negative and/or erroneous information, respond to it and copy your PIO as well as your colleagues, superiors, program director, and any other appropriate people.

Given the huge importance of Wikipedia as an information source, also consider asking one of their volunteer biographers to write your Wikipedia biography. The Wikipedia WikiProject Biography/Science and Academia Work Group site contains more information on this process. The site includes a list of contributors who write biographies.

Links to all these sites are listed in online resources for this chapter at *ExplainingResearch.com.*

Avoid Being Misquoted or Misinterpreted

Being misquoted or misinterpreted can be especially painful, given that as a researcher you prize precision of thought, speech, and writing. You will avoid much of that pain by taking steps outlined in previous chapters—including developing comprehensive news releases and practicing good interview techniques.

However, reporters will inevitably misquote you and/or misinterpret your work. And even if they quote you accurately, it may be out of context. Duke research communicator Joanna Downer advises scientists to accept that inevitability. "I tell them if they get upset about a quote, well, read it with your eyes crossed," she says. "Don't hang on every word and every comment. So they took eight words that you said in a ten-minute conversation and stuck it in quotes, and you are afraid that someone in L.A. is going to think you are an idiot. Do you believe everything you read? Do you believe that when you read that someone else was quoted about something...that is exactly what they said?"

As indicated previously, consider the media story only as a headline to alert audiences to your work. Make sure that your news releases and Web site include the whole story, to help offset the occasional misquote or misinterpretation.

Of course, you cannot allow flagrant misquoting or misinterpretation to stand. Biologist JoAnn Burkholder has been a victim of just such serious media misinformation, and her experience illustrates how to effectively correct it. In the 1990s, Burkholder and her colleagues at North Carolina State University reported evidence that fish kills in North Carolina bays arose from pollution-driven outbreaks of the toxic algae *Pfiesteria*. Her conclusions were vehemently denounced by local politicians, as well as the agricultural, seafood, and tourism industries. When she was also attacked by reporters who sympathized with those groups, she and the university responded assertively. In her essay "Uncertain Ground: The Boundary between Science and the Press," she describes one particularly egregious case:

> An individual from a small coastal newspaper became convinced by seafood industry lobbyists, environmental/health officials, and affiliated

> scientists that *Pfiesteria* was a non-issue. She launched a series in which I was the central focus of repeated extreme personal attacks. She also demanded written information from me, through our PI officers, about our *Pfiesteria* research, then deliberately and repeatedly published the opposite of what I had written, designed to cast me as incompetent. My university wrote a letter correcting her gross misinformation that the paper (which she also controlled) refused to publish. We posted the letter on our website. Recognizing the unprofessionalism of the individual, university higher officials and legal counsel dealt with all of her further demands and accusations. I was advised to respond to nothing she wrote because she had made it clear that she would not honor the truth nor allow any input from us. I followed that excellent counsel. It took two long years, but as our research continued to be vindicated by other laboratories, her unprofessionalism and misinformation were increasingly recognized.

Do Not Get Trapped on the Record

One hopes you never suffer the mental *ouch*! of being caught on the record when you thought you were off. Politicians and celebrities certainly remember acutely every gaffe they ever uttered when they thought the microphone was off or the reporter's pencil was pocketed. Protect yourself against such pain by always assuming you are on the record when you talk to a reporter and always assuming the camera and microphone are on when doing TV and radio interviews.

Also, remember that you are on the record even in social situations. University of Wisconsin PIO Terry Devitt offers a cautionary tale:

> We were hosting the CASW New Horizons in Science Briefing in Madison in 1995. One of our scientists—a very sophisticated, smart, experienced and, I thought, press-savvy guy—happened to be talking to Paul Raeburn, then the science editor of the Associated Press, over a drink about some of their exciting new research results. They had derived embryonic stem cells from a rhesus macaque—the first time such cells had been derived from a primate. Paul put down his drink and left the party to pursue the story. The results had not been published. I was put in the uncomfortable position of being asked by administrators and by this scientist who had spilled the beans to try to get Paul to not do the story. I reluctantly asked, and he declined. He said he was going to pursue the story, anyway.

Thus, recalls Devitt, the news of the advance was published in newspapers, threatening publication of results in major scientific journals. Importantly, Raeburn was by no means at fault. Like any journalist in such a position, he was perfectly within his rights to do the story, and in fact, his journalistic ethics demanded that he pursue it.

So, any time you are in the presence of a reporter, whether for an interview or socially, it is your responsibility to specify when you are going off the record or on deep background. Also, be acutely aware that any talk or other public event may be recorded and disseminated worldwide instantly in this age of live blogging, digital audio recording, and cell phone video.

Avoid a Credit Crunch

You can usually avoid accusations of hogging credit by following the guidelines in the chapters on citing colleagues in news releases and interviews. However, you might find yourself caught in a "credit crunch"—accused of overly emphasizing in news stories your role in a piece of work. In such cases, you can take extra steps to mollify aggrieved colleagues. Some examples:

- **Communicate directly through the publication.** Send a letter to the editor clarifying the role of colleagues who complain that they were overlooked in a news story. Even if the letter is not published, you have done your part to highlight that colleague's contribution.
- **Post your correction, "for the record."** Write an article or even produce a multimedia package for your Web site that covers the research and clarifies the colleague's role. Highlight that role by producing a video interview with the colleague. The article or multimedia package not only constitutes public acknowledgment of that role, but also will show up in search engine listings. Notify all appropriate webmasters of the Web package, so they can link to it. In particular, if the colleague is at another institution, let that institution's webmasters know.
- **Communicate with your colleague.** Send a message to the colleague recognizing his or her contribution and criticizing the media reports that neglected that contribution. Copy that message to all relevant administrators, including funding agency program officers, if appropriate.

Of course, in taking such steps, walk a fine line between correcting errors in credit and appearing to patronize your colleague. The best approach is to be perfectly factual in explaining the oversight and to avoid being too effusive and subjective in your communications.

Guard against Policy Foot-in-Mouth

You may also fall into the trap of commenting inappropriately on the policy implications of your research, especially if you are in a government laboratory. Even more complicated is the especially delicate situation of working in a government laboratory operated by a university.

In such a government-academic environment, academic freedom issues must be considered, says Catherine Foster, who was media relations manager at Argonne National Laboratory. "Our people think, and they can have opposing viewpoints to federal policy and should be allowed to express them. But they also need to understand that for the most part their money comes from a federal agency that may retaliate," she says.

If you want to express a viewpoint, yet protect yourself against policy foot-in-mouth, first educate yourself thoroughly not only about the policy issue, but the political machinery underlying that policy. As discussed previously, your best allies are the government relations officials in your institution or professional society. And once you decide to speak out, you can work with those experts, as well as your PIO, to craft messages that minimize your risk and maximize your message.

Disclose Corporate Ties and Other Influences

You will preserve your credibility and make your life far easier if you make emphatically clear any corporate funding sources or other possible influences on your research. Even an inadvertent lack of clarity can come back to bite you. For one thing, journalists will find out, and the resulting story will be at the least embarrassing and perhaps career-damaging.

A classic case was the controversy surrounding a 2006 paper in the *New England Journal of Medicine* by Weill Cornell Medical College physicians Claudia Henschke and David Yankelevitz. They reported evidence that CT scans could prevent a large percentage of lung cancer deaths. In 2008, the *New York Times* and the *Cancer Letter* disclosed that the study was funded by a foundation supported almost entirely by the parent company of the cigarette maker the Liggett Group. Henschke was president of the foundation, and officers included colleagues and medical college administrators.

Even though Henschke and her colleagues declared that they had no intention to hide their funding source and had been open about the foundation, critics thought otherwise. "If you're using blood money, you need to tell people you're using blood money," the *New York Times* quoted American Cancer Society chief medical officer Otis Brawley as charging. What's more, failure to adequately explain her funding source damaged her efforts to promote CT screening for lung cancer. The *New York Times* quoted lung cancer expert Paul Bunn as saying, "She's the biggest advocate for widespread spiral CT screening.... And now her research is tainted." Such lack of disclosure, whether purposeful or an inadvertent communications failure, compromised the research. And, such damage could cost lives, for example, if it turns out that CT scans can catch lung cancer at early stages.

Beyond stating your funding sources, also disclose your consulting agreements, stakes in companies, and even paid speaking engagements. Make such arrangements clear in news releases and on your Web site, because the media will seldom detail them in news stories.

Also go beyond any basic reporting requirements, such as those required by journals. PIO Don Gibbons of the California Institute for Regenerative Medicine offers the cautionary story of Harvard scientist JoAnn Manson, an expert on the weight-reduction drug Redux. In a 1996 editorial in the *New England Journal of Medicine*, Manson and Gerald Faich commented on a previous *NEJM* article warning of a dangerous, but relatively rare, side effect of Redux. In their editorial, Manson and Faich asserted that Redux's benefits may outweigh such risks. A credibility problem arose because when she submitted *NEJM*'s conflict-of-interest disclosure form, Manson did not mention that she had briefly served as a paid consultant to the company that made Redux.

The misunderstanding received widespread adverse publicity, even though, as *NEJM* editor Marcia Angell noted, the required form could have been misleading. It asked authors to disclose "ongoing financial associations," or "regular consultancies," which Manson interpreted to not require disclosure of brief consulting stints. "She adhered to the letter of the law but not the intent," says Gibbons. "It's important to disclose absolutely everything, because black eyes do not go away quickly."

Do Not Skew Your Perspective

More broadly, do not let corporate ties bias your research explanations. Reporters are well aware of such machinations, and the benefits they offer researchers. Trudy Lieberman described the situation in a 2005 article, "Bitter Pill," in the *Columbia Journalism Review*:

> Scientists who test the drugs tend to talk up the product's strengths to the press. "It's not that scientists lie, but if they say certain things,

> they get rewarded," says Dr. Bruce Psaty, a professor of medicine and epidemiology at the University of Washington. If these experts speak favorably about a drug to the press, they tend to get invitations to speak about the drug at conventions for doctors and at educational seminars that hospitals offer for their employees, where they get a chance to further promote their study results. All these activities help enhance careers and bring good press to the clinic or the university. "It used to be death to get your name in the paper if you were an academic," says Sherrie Kaplan, associate dean of the College of Medicine at the University of California at Irvine. "Now academics are elbowing each other to get on the *Today* show." For those who learn how to market their studies, the visibility brings the next round of grant money. The more Congress and the public hear about the study, the more potential support from the National Institutes of Health.... The more other drug makers hear about a scientist's study, the more likely they are to seek out him or her for the next clinical trial. And the more likely journalists are to use that scientist again.

Escape Attacks from Vampire Speculation

Assume that any speculation you advance in public—even one you consider whimsical—can become a "vampire speculation" that will not die. Journalist/author Keay Davidson cites the cautionary case of University of British Columbia biochemist David Dolphin. At a 1985 AAAS meeting, Dolphin speculated that mythical werewolves and vampires may have been people who suffered from the disease porphyria. Davidson recalls what followed. "The media just went nuts over this goofy little story based on perfectly legitimate research, and when I called him years later, he said 'Every Halloween I get flooded with calls. It has ruined my life.'" Thus, Davidson warns "a scientist to think long and hard particularly if what you are proposing is something that is pretty cutting edge, or is highly speculative and concerns a rather lurid or sensational topic; something that you are not absolutely certain of. You won't win any kudos from your colleagues for being in the press."

Cope with a Controversy/Crisis

While you might see a controversy coming and prepare for it, a controversy might also explode in your face, requiring a rapid and unrehearsed response. The case of the "gay sheep" story offers a prime example of such unpleasant surprises and how to manage them.

In 2004, the Oregon Health and Science University issued a news release on what they thought would be a newsworthy, but not particularly controversial, research finding. Charles Roselli and colleagues reported in the journal *Endocrinology* that they had identified a biological basis for male sheep's preference for same-sex partners. The university news release quoted Roselli as saying, "This particular study, along with others, strongly suggests that sexual preference is biologically determined in animals, and possibly in humans. The hope is that the study of these brain differences will provide clues to the processes involved in the development and regulation of heterosexual, as well as homosexual, behavior."

To the university's and researchers' surprise, the release triggered a storm of protest from animal rights activists and gay advocates, based on a misinterpretation of the work—namely, their belief that the scientists were attempting to "cure" the homosexual rams with hormone treatments. One source of the misinterpretation could have been the following sentence from the university news release: "[The scientists] would also like to know whether sexual preferences can be altered by manipulating the prenatal hormone environment, such as by using drugs to prevent the actions of androgen in the fetal sheep brain."

In protest, People for the Ethical Treatment of Animals (PETA) implored its members to send a message to university officials denouncing "cutting open and killing gay sheep in an attempt to alter sexual preferences in animals." In an article in the *New York Times* by John Schwartz, PETA representative Shalin Gala cited a quote in an earlier news release about the work as implying that the aim of the research was to *control* homosexuality. The release quoted Roselli as saying that the research "also has broader implications for understanding the development and control of sexual motivation and mate selection across mammalian species, including humans." Jim Newman, the PIO who wrote the release, replied that the word "control" was used in the scientific sense of understanding how the body controls processes, not in the sense of exerting control. And Roselli defended himself by asserting that merely mentioning possible human implications of the work did not mean that any human application was intended.

Nevertheless, the news release's use of the word "control" and the mention of human implications left both the PIO and the scientist vulnerable to the controversy.

Bioethicist Paul Root Wolpe agreed, asserting in the *Times* article that "I'm not sure I would let him [Roselli] off the hook quite as easily as he wants to be let off the hook," and that by discussing human implications of the work, Roselli had "opened the door" to the reaction and "has to take responsibility for the public response."

The Roselli case offers two object lessons in handling potential controversy: First, try to anticipate how your work might be misinterpreted and take into account those possibilities in your communications. Second, take aggressive action to correct such misinterpretations, as the researchers and university did. They responded to every e-mail message, issued clarifying statements, and commented on blogs that posted erroneous information.

With the cautionary "gay sheep" story in mind, here are principles to help you cope with crisis and controversy:

Consult Your Communicators Early

- **Ask your PIO to help protect you against future controversy by laying a communication foundation well before any problem rears its head.** This foundation should include news releases, feature articles, and Web content that constitute vetted information on your research for media and other audiences. What's more, the working relationships you have developed with communicators give them a head start in understanding any problem and helping you cope with it.
- **The instant you see trouble brewing in your research, call in your communicators.** Involve them well before the feces strike the fan—if possible, even before there is a fan. Even involve your communicators in preparing reports to official bodies such as regulatory agencies about the problem. Such reports invariably become public, and your communicators can help you word those reports so that they read to your advantage when quoted in media articles.
- **Tell your communicators everything, even though there may be some aspects of the problem that are painful.** The State of Denial is a lousy place in which to get trapped.

Once you call in your communicators, *listen to them.* They may tell you things you do not want to hear, but they do so out of their sense of duty to the institution and to you. However, their advice might not be forthcoming because of their junior position in the organization, says Ohio State PIO Earle Holland. Given that communicators may be lower level administrators, they might not be comfortable telling truth to power, he says:

> It is very tough to sit there across the table from the president of your university and say "No, you cannot do that. If you do that, this is what is going to happen."… Somebody at our level, even a senior-level research communicator, saying that kind of thing to the senior leadership is not

often well received. The chances we take when we do that are enormous, but the chances we take by not doing it are so much greater to the institution.

Disclosure: Rip Off the Bandage

Most crises will require you to disclose painful information. Good communicators will advise you to tell everything, and tell it quick. Take that advice. Ripping off the disclosure bandage offers several advantages. For one, it shortens the pain, turning what might be a multiday media story into a single-day story. Also, full disclosure gives you credibility. You will not be able to keep the information secret, anyway. Any statement, e-mail message, or video, even though labeled confidential, will likely find its way to the public.

Keay Davidson offers an example of how slow peeling of the disclosure bandage proved excruciating for an institution and its PIO. The occasion was a story he was covering for the *San Francisco Chronicle* in 2006 about exposure to plutonium of workers at Lawrence Livermore Laboratory. The laboratory's PIO, Susan Houghton, declined to discuss the case in detail, stating that it strictly involved the contractor and its employees. The lab was not responsible for the incidents in any way, she insisted. When Davidson published a story stating that workers had been exposed, "she called my editor and yelled at him, telling him that I was trying to damage the lab, and I was biased," recalls Davidson. But the incident also showed the newspaper's biases, notes Davidson. "The *Chron* is a very white, very conservative newspaper, not given to investigating major local figures or institutions (except black baseball players like Barry Bonds and black governmental officials and gangsters in Oakland), and my editor, Terry Robertson, a normally amiable sort of guy, sympathized with her," recalls Davidson. "Then he came back and lit into me. Over the years, the *Chron* almost always buried my multiple 'hit pieces' on Livermore deep inside the paper."

Some weeks later, Houghton "called the *Chronicle* to apologize and to say she had erred," says Davidson. "She admitted that two Livermore employees had been contaminated with plutonium, and that she had not known this earlier." Also, says Davidson, three contractor employees were contaminated. Recalling the case, Davidson says, "I can't prove that she was brazenly lying to me originally and then got cold feet, and I can't prove that she was so inept that she was unable to beat the truth out of the people she works for. But I know one thing: it was her job to get the truth for me, or at least to say 'I don't know. I will try to find out.'"

Obstructionist administrators and PIOs might believe it necessary to withhold information to "protect" their institution. However, such so-called protection damages the long-term credibility of the researcher, the institution, its administrators, and its communicators, says Davidson. "They have a special job to be frank and honest, and don't just B.S. the press, because once you wreck your credibility with us, it is very hard to get it back," he says. "And trust me, there'll always be one self-righteous, closeted left-wing prick—some journalist like me, who grew up dreaming of becoming [pioneering investigative reporter] I.F. Stone but never did—who will be so mad at you for the deception that he'll do exactly what I did after that: mention Houghton's screwup in every story possible!"

Misguided administrators may also cite the "good of the public" as an excuse for not releasing controversial information. This ploy more often is merely a cover for their timidity about facing the media and the public. Truly serving the good of the public and the institution almost always means full, immediate disclosure of information. Far more often, the lack of information only breeds suspicion and distrust. And more practically, the media will discover all the information, anyway.

As an example of serving the public good, Rodgers offers the debate that took place at Johns Hopkins over disclosing that one of its breast surgeons was HIV positive. Some administrators were reluctant to announce the information, she recalls:

> They took a paternalistic approach, well known among bioethicists as a two-edged sword, that '"If we told the four thousand women on whom he operated over his lifetime, they will become, however inappropriately, terrified, and possibly blame or sue the institution."' Our PIO team pointed out the fact that this couldn't be kept secret; some people knew and rumors were flying. We had had one press inquiry about it.
>
> We further pointed out that we were going to be on the defensive if the situation was verified by a third party. And most of all, we emphasized that as a leading academic medical center, we rigorously required universal precautions to prevent transmission and had an obligation to address inappropriate fear and fear-mongering when the facts failed to support either.

Such enlightened arguments carried the day. The university, with permission from the surgeon's family, disclosed that the surgeon had HIV and sent letters to all of his patients. It participated in a look-back study with the state health department to determine the health status of the surgeon's thousands of patients, and it subsequently published the results, which reaffirmed the very low risk to patients in such situations. Not one patient was HIV-positive; there was no mass hysteria or demands for mass testing. "It taught us that, as a good scientific communicator, if you follow the scientific process, you will get the best outcome for patients and the institution. This is a classic case of doing the right thing." says Rodgers.

A good tactic for cushioning the blow of disclosing a problem is to include information about the solution when releasing information about the problem. The result will be a more positive story. So in planning to disclose a problem, quickly develop at least an idea of the remedies you plan, so that you can discuss them when you reveal information on the problem. For example, when Hopkins released information on their HIV-positive surgeon, they also announced their plan to communicate with his patients and to fully study and report on any clinical impact.

Holland warns, though, that in discussing a problem, you should be wary about being lured outside your factual comfort zone by media questions. He tells researchers in such situations "I want your answers to be directly linked to your expertise and your involvement in this. I don't want you to be drawn into a realm of 'What do you think about this?' or giving your own personal feelings about an ethical issue."

To help you avoid saying the wrong thing during a controversy, conduct a "murder session," advises Friedmann. Such sessions are intense Q&As, in which a hostile questioner drills you with questions designed to probe and correct weaknesses in your answers. Your PIO might bring in an outside consultant trained in managing controversy to conduct such a session, notes Friedmann. "Even though scientists intellectually understand it's role-playing, don't discount the psychological damage this exercise can have on working relationships," she says. "Better to have someone else slash them to ribbons."

Finally, do not play favorites among journalists when ripping off the disclosure bandage. You might be tempted to give preference to "friendly" reporters, but you will only alienate the majority of journalists by doing so.

"No Comment" Is Really a Comment

You might be tempted to issue a terse "no comment" during a controversy. The problem is that when your "no comment" appears in the media, it portrays you as stonewalling and suggests there may be something fishy going on. So, if for some reason you really cannot comment, make your statement a "no comment, *because*..." Explain as completely and honestly as possible why you cannot comment at this time. Tell about any legalities, regulations, policies, or ethics that prevent comment.

Joe Palca offers as a good example of the damning effect of a "no comment" how a PIO with NIH refused to respond to an inflammatory animal research video obtained by PETA. The video showed researchers laughing as they used a hammer to remove an electrode from the head of a monkey. Palca, then a television assignment editor, sought comment from NIH, which had funded the research. The PIO asserted that NIH would have no comment.

Palca said to the NIH PIO, "You're joking, right? You funded the research, and it's your responsibility to present the arguments for animals in research and to have somebody stand up and say 'Look, this is an aberration.'" Still the NIH PIO refused to comment. Fortunately, says Palca, a senior medical researcher was willing to publicly denounce the behavior depicted on the tape. What's more, the researcher used the chance to defend responsible animal research as crucial for developing new drugs and treatments.

Be Ready to Go Real-Time

Sometimes events move so quickly that you will find yourself in a real-time-response mode. Try your best to operate on a full-disclosure basis, even as events unfold. You might need to go back and correct yourself, but the media and the public will be much more forgiving. An excellent example of real-time crisis response was the notorious 1989 case of the unleashing of an Internet-clogging computer worm by Cornell student Robert Tappan Morris. As Cornell science editor at the time, I helped manage the crisis.

On a Friday evening I received a call from then-*Washington Post* science writer Philip Hilts telling me that technicians had traced the computer worm wreaking havoc on the Internet to Morris, a Cornell graduate student. Hilts wanted to find out what we knew. We knew nothing then, but I promised to get back to him as quickly as possible. I immediately called the chairman of Cornell's computer science department and informed him of the situation. He knew nothing about the worm, either. I told him that the *Post* had the story and that the best way to show Cornell's cooperation was to tell Hilts everything we were discovering, even as we discovered it.

The chairman immediately directed the system administrator to sift through Morris's files. The system administrator quickly discovered Morris's computer code for the worm, and I called Hilts and put the department chairman on the phone, who disclosed everything they were finding. The story exploded in the media, but at no time during the controversy was Cornell accused of any complicity in the worm or of attempting to hide the facts. Cornell continued its policy of openness by preparing and releasing a thorough report on the events.

There may be times when you are really not prepared to talk when you receive a call from a reporter who has exclusively ferreted out a controversial story. In such a case, you can often buy time to marshal your information by striking a bargain. Offer to cooperate fully and to allow the reporter to break the story, in return for the reporter's promise to hold off for a reasonable time. Thus, you have rewarded the reporter's investigative efforts and made sure that the reporter has all the facts. But you also have given yourself time to gather those facts and develop a plan to correct the problem. And, you have cultivated good relations with the reporter, who will no doubt be covering the controversy further. Conversely, if you react to such a reporter's call by prematurely giving the information to all the media, you will alienate the reporter and possibly release a half-baked story that does not tell your side fully.

Understand That Emotions May Trump Facts

Emotion can sometimes trump facts during a controversy. It may not matter that, as in the Roselli "gay sheep" case, the controversy was triggered by a misinterpretation of the work. Your communications during crisis must take that emotion into account.

"Trying to teach scientists the difference between facts and emotions is very difficult," notes Rodgers. "They make their blunders because they want to stick to facts, when in fact the story has an emotional or social component." Rodgers cites as an example an NIH consensus-development conference on mammography. The conference found that existing data clearly indicated that women under 50 did not benefit from annual mammograms. Such tests would find only a minimal number of cases, compared to the risks from unnecessary biopsies, for example. At the news conference announcing their conclusion, the scientists—overconfident about the power of the facts—did not take the emotional factor into account, says Rodgers:

> They knew that in the audience were people representing breast cancer advocacy groups, women's groups, foundations devoted to breast cancer. Questions surfaced for the scientists such as "If this were your wife, would you feel the same way?"
>
> One panel member said he would leave it up to any woman worried enough about it to make a decision right for herself, which contradicted in some way the overall public health approach of their policy recommendation. Another panel member suggested such questions were not the point. But of course they were the point for some members of the public and press.

Ultimately, the issue was as much a political as a scientific one, and despite the data, the recommendations were overturned by NCI leadership, recalls Rodgers. "It's always well to remember there is a human being attached to the other end of a breast," she says. With even the most basic science, from cosmology, to the virology of HIV, to breast cancer research, says Rodgers, "there is going to be an emotional human component, a reaction to information. You have to anticipate that and address it upfront when you communicate."

So, consider preparing an "emotional spreadsheet" in coping with a controversy or crisis. List all the involved groups, what emotions they bring to the issue, why they feel those emotions, and how your communications can deal with those emotions. Include yourself in that emotional spreadsheet. Are there emotions or biases you bring to the controversy that may cloud your judgment and lead to unwise actions?

There have been cases in which such emotions have driven researchers incensed about a media story to rashly call to threaten the reporter with a lawsuit. Such stories were often perfectly accurate, and it was only the emotion of the moment that triggered such rash action. But even if there were errors in a story, more reasoned approaches, such as asking for corrections, would have led to a better outcome.

Consider the Preemptive Strike

If you know that a controversy is looming, consider a preemptive communication strike to ensure that your point of view sets the tone of the debate. One good example of such a preemptive strike is the case of communication about release of the genetically engineered virus at Cornell, discussed above.

Another good example was the university that found itself facing an upcoming protest rally against its animal research. The dramatic TV-newsworthy rally would certainly garner major publicity. When a university PIO asked me for advice, I suggested a preemptive strike. So, the day before the protest, the university called a news conference at which they explained their animal research and highlighted its medical benefits. The news conference included personal testimonials from patients whose lives had been saved by drugs and medical procedures developed using animal models. As a result, the university grabbed the initiative, and the animal rights protest received less coverage as a second-day reactive story.

Do Not Tiptoe around a Delicate Story

Then there is the story that does not reflect badly on your institution but contains elements that are "delicate." You might be tempted to tiptoe around delicate aspects in order to spare the sensitivities of the media and the public. If such tiptoeing compromises your information, resist that temptation.

Rodgers cites the example of a release Hopkins was preparing on a 2007 *New England Journal of Medicine* paper showing that oral sex increases the risk of throat cancer linked to human papillomavirus infection. A less-than-forthright initial draft of the news release failed to communicate the work clearly, because it tiptoed around information about sex practices, says Rodgers. "There was no mention of fellatio, cunnilingus, or what oral sex was," she says. "And you didn't encounter the phrase 'oral sex' until the ninth paragraph. However, cooler heads prevailed, and we rewrote the release to state the research plainly."

Thus, the rewritten release contained the crystal-clear declaration "Oral sex, including both fellatio and cunnilingus, is the main mode of transit for oral HPV infection, the investigators say, although mouth-to-mouth transmission remains possible and was not ruled out by the current study." Ironically, despite such frankness by the university, and the clear public health implications, many newspapers did not carry the story, recalls Rodgers. "The *Washington Post* covered it, but the *Baltimore Sun* didn't, nor did any television network. We got calls from a newspaper in Texas saying 'We can't use words like that.'"

Rodgers warns against allowing such media prudishness to stop universities from writing about any aspect of science. "It can become a vicious circle, where the media won't cover something, so the scientists become reluctant to talk about it, so PIOs self-censor. So, part of our job now as PIOs is not only to persuade the press they have to cover something; it is to persuade the scientist that the press ought to cover it," she says.

Prepare for a Hyperstory

Watching an onrushing "hyperstory"—for example, a research achievement sure to attract vast media coverage—can be like standing before a tidal wave: you can be inundated by it or, by preparing, you can surf it. A key to riding the hyperstory wave is extensive planning and investing in communications.

A textbook case of wise planning for a hyperstory was the University of Wisconsin's preparation for the 1998 announcement of the first culturing of human embryonic stem cells. PIO Terry Devitt and his colleagues worked for six months on the communications. They developed a comprehensive, continually updated Web site with the simple URL *StemCells.wisc.edu*. They created a professional-quality graphic depicting the process. They media-trained researcher James Thomson and others involved. "Also very important, I was fortunate to have really good administrators, really good administrative support," says Devitt. "The dean of the graduate school, who is the chief research officer at Wisconsin, was incredibly supportive and understood what we had to do. She gave us the backing that we needed to do our jobs well and to represent the institution and to cope with the huge controversy." This support included allowing Devitt to attend meetings of the bioethics and other committees overseeing the research.

When the researchers announced the achievement in the journal *Science*, Devitt and Thomson had prepared for the media onslaught, says Devitt: "Thomson sat in the office next to mine for a week-and-a-half and did nothing but take phone calls," he recalls. "My phone didn't stop ringing for two weeks. I would take calls; I would do triage; I would make a list; I would go next door and hand him the list of the calls to be returned. He probably did about a hundred interviews."

Finally, even after a crisis or hyperstory has passed, your communications efforts should not end. Continue to issue news releases and update and enhance Web sites, videos, feature articles, and other content. They are an information gift that keeps on giving—to your research and your reputation.

26

Manage Media Relations at Scientific Meetings

While major scientific societies set up and run professional media operations such as press rooms during their meetings, you might be called on to manage the media operation at a smaller conference or symposium. This chapter will help you effectively communicate your meeting's news to media.

First and foremost, arm yourself with knowledge. Learn how the pros do it by visiting the press rooms at meetings such as the American Association for the Advancement of Science, American Astronomical Society, American Cancer Society, American Chemical Society, American Geophysical Union, American Heart Association, American Medical Association, American Physical Society, and Society for Neuroscience. Observe how the PIOs at those meetings develop and manage their media operation.

Even though as a researcher you might have no experience managing a media operation, with some preparation you can do an excellent job of attracting media coverage. A premier example of how one committed scientist can make a major difference in coverage of his society's meetings is the case of Stephen Maran, an astronomer who became press officer for the American Astronomical Society in 1984. "When he started to get involved, he made the effort to really learn and understand how the media operated and what reporters needed," recalls *Science News* editor Tom Siegfried. "He made sure he knew every reporter's name, and who they were, and what they wanted, and established a system where communication was ongoing, and ran press rooms designed to meet the needs of the reporters. One person made astronomy coverage vastly better than it would have been if they had hired ten of the best PR agencies in the universe."

Identify Newsworthy Papers

Your first step is to identify newsworthy papers for publicity. Scan the conference program and ask meeting organizers what they believe are the most scientifically significant papers. Then e-mail the authors suggesting that they contact their PIOs about doing news releases, to be embargoed to their conference presentation.

For larger meetings, send the abstracts of talks to the PIOs at authors' institutions and let them decide whether to do a news release. You can identify PIOs at those institutions by accessing the AAAS database *Science Sources* through *EurekAlert.org*. To access Science Sources and to submit news releases to EurekAlert!, you can request to register as a PIO for your society.

Of course, some researchers might decline to do news releases because they understandably want to preserve their ability to publish in major journals by not publicizing their results. In such cases, even though media will have access to the conference program, you should not highlight the paper as newsworthy. Such highlighting constitutes the proactive publicity that can disqualify a paper from publication in journals such as *Science* or *Nature*. However, do not attempt to hide those abstracts from reporters. They should have access to the full program.

Notify the Media

To alert media to your conference, post a meeting notice on the calendars of the major online research news sites: AlphaGalileo (Europe), Ascribe, EurekAlert!, Newswise, PR Newswire, and ResearchSEA (Asia).

That meeting notice should link back to your conference Web site, which should include abstracts of papers scheduled for the conference. Although reporters may prefer to cover your conference without physically attending, encourage them to actually attend the meeting, advises Maran. "There is an enormous advantage to getting a newspaper reporter to your meeting," he says. "If [as a reporter] your editor has sent you to Honolulu for our meeting, you are going to file." Attending reporters will typically file one or two stories a day from a meeting, says Maran. However, still offer full information services such as news releases to those reporters who do not attend, covering your conference from their home base.

The conference Web site should include an online newsroom, which can greatly encourage press attendance and coverage. This online newsroom should include information on press registration and press contacts, collections of abstracts, a press conference schedule, information on embargoes, and an explanation of the facilities available to reporters. You should also create an online press packet that contains embargoed news releases on papers, the press conference schedule, and links to photos, illustrations, and videos submitted by PIOs. Importantly, this press packet should *not* be posted on a public Web site, but only in the embargoed areas of EurekAlert!, Newswise, or other research news sites. Public posting of such materials breaks the embargo, allowing reporters to write about the papers even before they are delivered.

To attract reporters to the conference, also compile a list of feature story ideas that reporters can do by interviewing conference attendees. Your conference organizers and colleagues can help develop such ideas. Conference news releases and feature ideas will not only spark coverage of the meeting itself, but long-term coverage of your discipline. For example, says Maran of the materials he distributes at the American Astronomical Society meetings, "Reporters lap them up, and the main target all along has been not even reporters but editors at popular astronomy magazines around the world. And they have almost no budget for articles, and the press releases from our meetings become stories, and in some cases just pictures with a long caption in magazines, for weeks and months to come."

Organize a Newsroom

The ideal news operation at the meeting will include workspaces for reporters, display space for news releases and papers, rooms for interviews and news conferences, and a social room. Of course, for a small symposium you might have only a couple of meeting rooms for reporters. Your "news release and paper room" might be only a table outside the press room and the "social room" the hotel bar. In any case, even a small dedicated space for reporters is useful.

Larger meetings also offer reporters WiFi Internet connections, phone, fax, and other resources, although for a smaller meeting the hotel business center might have to suffice.

Set Embargoes on Presentations

Set as the day and time of delivery the embargo on media stories on meeting presentations. However, schedule news conferences covering presentations well before the presentation. For example, for presentations in the morning, hold the news conference the afternoon before; and for presentations in the afternoon, hold the news conferences that morning. This scheduling enables reporters to write their stories to be released right when the session ends. Reporters will honor such embargoes, because they are in the reporters' best interest and because you can threaten to exclude them from receiving materials on future meetings if they do not honor them.

When presentations are delivered in an open meeting attended by reporters, they are fair game for news articles, live blogging, and audio and video recording. Sometimes meeting organizers have attempted to dissuade reporters from such coverage, for example if they might compromise scientific publication. Do not try to prevent such communication, or you will alienate reporters and PIOs and possibly affect future coverage of your conference. It is up to the speakers to understand the implications of media coverage for their scientific publication and to take steps to guard that publication. There are cases, however, in which it might make sense to ask reporters to hold any coverage until the end of a conference. One such case is if the conference will result in a consensus report. For example, at the historic 1975 Asilomar Conference to discuss safety of biotechnology, the reporters agreed to hold their articles until the conference was over. "None of us was experienced or knowledgeable enough to understand everything," recalls David Perlman, who covered the conference for the *San Francisco Chronicle*. "We all had agreed to not write or file anything until the meeting was over. And at the end, they were going to have ... [biologists] Paul Berg and Sydney Brenner explain what it was all about." As a result, says Perlman, coverage of one of the most critical meetings in the history of biotechnology was more informed and complete. In any case, explicitly state on the conference Web site the ground rules regarding media coverage, including live blogging and recording of talks.

Plan and Conduct News Conferences

A news conference is an ideal venue for reporters to learn about your conference's research findings or symposium topics. The news conference enables reporters to ask their questions without competing with scientists in the conference session. And they can benefit from their fellow reporters' questions and the answers to them. Even if you are a researcher with no experience in news conferences, following the guidelines in this chapter will enable you to conduct perfectly satisfactory sessions with reporters. However, if you do not have the time, or you feel uncomfortable in the PIO role, consider hiring a consultant who is expert in organizing and conducting scientific news conferences.

To plan your news conferences, select the symposium sessions that look most newsworthy and invite their chairpersons and a few key speakers to participate in news conferences. Such sessions typically involve no more than five people, with three the ideal number. Invite participants well in advance of the meeting, so they can prepare news releases, visuals, and other materials. Suggest that participants, especially those who may be controversial, prepare a statement beforehand to read, so that they can explain their work the way they wish to, rather than counting on the right questions from reporters.

Once participants have accepted, notify their PIOs and suggest that the researcher and PIO conduct a dry run of their news conference comments, to make sure they communicate their work clearly. Also, cautions Maran, make sure their abstract actually reflects the paper. "In some cases they have not yet obtained the result described in their abstract, but expect to have it at the meeting," he says. "There is not much harm done, unless you've announced they are going to give a press conference. The most common case is that what was described in the abstract submitted months earlier just hasn't panned out."

Set Up Remote Access

Reporters who do not attend the meeting can cover the news conference remotely if you make it available via a telephone conference call. Work with an audio/visual technician to connect the room's sound system to a phone-in line, preferably a toll-free number. Thus, callers will hear the speakers clearly, and their questions will be heard over the sound system. Have participants speak into microphones rather than a speakerphone, because it will pick up room noise, and the sound quality will be too low for reporters who want to record the conference. Also, their questions will not be clear.

To enable reporters to see slides projected at the news conference, your facility's AV technician can set up a "slidecasting" system. This system sends any visual shown on the computer projector to a Web site that remote reporters can access to see the slides as they are discussed. For very special news conferences and if you have the budget, broadcast them via streaming video via the webinar systems described in chapter 15.

Prepare the Participants

Before the news conference, meet with participants in a quiet room to go over details. Science communication consultant Lynne Friedmann—who has conducted more than 300 news conferences during a decade of AAAS annual meetings—offers these steps for preparing the participants:

- **Determine the order of speakers.**
- **Explain that you will recognize reporters to ask questions.**
- **Tell them they need to initially summarize their findings completely, succinctly, and at a lay level—putting their findings up front—after which you will open the floor to questions.**
- **Brief them on what will be available at the head table: water, notepads, pens, and so on.**
- **Tell them to leave paraphernalia such as coats and briefcases in their hotel room or in a prebriefing room you will provide.**
- **Make sure they understand the etiquette of using microphones with media.** For example, radio and television reporters may use the microphones for recording, so participants should not make extraneous noise. Also, such recording means they are "on the air" at all times, so their side comments may be captured.
- **Tell them that you will use signals to keep the conference moving, such as standing up or making eye contact when they are going on too long.**

Friedmann advises warning participants about two particular pitfalls: First, they should not assume that the news conference is a teaser to attract reporters to the main session. "I tell participants that many of the reporters do not have the luxury of going to your three-hour session tomorrow," she says. "They will be writing their story from this forty-five minute briefing. If you have conclusions, if you have data, this is an appropriate forum. This is not the time to keep your cards close to your chest."

Second, make participants understand that they need to explain their work memorably, using vivid metaphors and all the other techniques described in this book. Even Nobelists may not grasp this concept, says veteran science reporter Robert Cooke. "Lars Onsager, who won the Nobel Prize for the chemistry of irreversible reactions, had a news conference, and he couldn't explain what the hell he meant by an irreversible reaction," recalls Cooke. "[Caltech biologist] James Bonner stood up and said, 'Well a good example of an irreversible reaction is you can't unfry an egg,' which was perfect; everybody understood that."

Conduct the News Conference

For the news conference, wear a formal businesslike suit or dress. However, Friedmann advises adding a brightly colored element such as a tie or scarf that makes you stand out, so reporters will remember who you are. Also, carry a clipboard, both to organize papers and to establish yourself as the conference official.

After the opening statements, during the question period, "Keep the pencils moving," advises Maran. "The way I judge how well a news conference is going is how many pencils are moving—or nowadays how many fingers are typing at laptops—versus how many people are sitting and looking out into space," he says. When Maran sees too much staring, he will curtail a speaker and move to the next question.

If an argument breaks out between a reporter and a researcher, be ready to intervene. If you believe either party is preventing the other from having a say, ask the other to let that party speak. If the argument is not helpful to the rest of the reporters, suggest that it be pursued privately. Similarly, if a reporter's questions are becoming too arcane or detailed, suggest that the discussion be continued after the news conference.

Try to scout problematical news conference attendees ahead of time, in order to avoid disruptions. For example, Friedmann recalls learning that a hostile writer planned to attend an AAAS meeting news conference with chemists Sherwood Rowland and Mario Molina, who study atmospheric ozone. "He did not believe their findings and went to every press conference and public forum in which they presented, solely to bait them into an argument," recalls Friedmann. "Knowing who that individual was, I was able to alert the scientists to his presence. During the Q&A, I called on every writer but him, but eventually there were no other hands up. At that point he was recognized, but by then it was the end of the press conference, and his inflammatory remarks were cut short without the desired effect."

Celebrity scientists may present particular challenges, says Friedmann—not because of the celebrities themselves, but because of the star-struck people around them. She recalls, for example, the case of an AAAS meeting news conference with author/physician Michael Crichton:

> An AAAS officer who heretofore had nothing to do with the briefing pulled rank on me and told me he was going to be responsible for escorting Michael Crichton out of the press conference when it was over. I remember thinking to myself "Good luck." After the briefing, I sat there and watched as this pompous little artichoke was jumping up and down, saying "Make way," and none of the reporters knew him from Adam, and they were mobbing Crichton.
>
> Finally at one point the guy from AAAS comes stomping over to me and says, "It's your room, you get him out of here." Without a word, I walked up behind Crichton, reporters looked up at me, I tapped my

> wristwatch, they respectfully stepped back, and I got Crichton and walked him out of the room past this guy. Point, set, match!

The three key tips for handling celebrities, says Friedmann: establish your authority, exert it when you need to, and plan an escape route.

Finally, end the news conference after 45 minutes to an hour. However, do not be bashful about cutting it off at any time that the questions start to flag or reporters seem to be ready to leave. They can always ask follow-up questions individually.

Arrange Interviews and Make Experts Available

As conference press officer, you also will be asked to arrange individual interviews and provide experts for background discussions. So, before the meeting, make sure you know how to contact any participant in whom reporters might be interested. Obtain their cell phone numbers and email addresses, and check whether they are willing to talk to reporters.

Background experts are important because reporters prefer to seek a second opinion on stories about research findings. These experts, in fact, need not be attending the meeting, although it is best that they are. Science communicator Cathy Yarbrough developed a particularly effective system at the American Heart Association meetings to have such experts constantly available. "I had these scientists lined up to work on call for specific periods," she says. "Also, I had a special room for them called the 'spokespersons room.' It had telephones, couches, and food." In fact, says Yarbrough, good food was her secret weapon to lure reporters and scientists to the press room. "The scientists we recruited wanted to come down and eat with the reporters so they could get a good meal and not have to battle everyone at the convention center," she says.

27

Should You Be a Public Scientist?

Now that you understand the tools to explain your research, a central question for your career is how extensively you want to use them. That is, where do you position yourself on the "private-public" spectrum—ranging from the "private scientist," who communicates only to your colleagues, to the "public scientist," who uses communication tools and the media spotlight to highlight the issues important to you?

To help you make that decision, recall some of the key questions posed in chapter 2:

- Why do you want to explain your research?
- Are you a natural explainer?
- What vehicles do you prefer to use to explain your research?
- How much time and effort are you willing to devote to explaining your research to lay audiences?
- Does your research field require explaining?

After reading this book, you should have a good idea of the answers to these questions and how those answers influence your communication strategy. To give you more help in deciding about your place on the communication spectrum, this chapter explores the benefits and costs of being a public scientist.

Throughout this book I show that communicating your work to broad audiences is an integral part of your profession. As John Ziman was quoted in the introduction, "The objective of Science is not just to acquire information nor to utter all non-contradictory notions; its goal is a *consensus* of rational opinion over the widest possible field." Public scientists, however, go beyond explaining their research just to advance their own work. They overcome a natural preference for the laboratory and the seminar room to lead the public debates on science- and technology-related issues.

They recognize that scientist-educators bring invaluable teaching skills to these issues and can counter the voices of pseudoscientists and their representatives. For example, as biologist/filmmaker Randy Olson warns in his film *Flock of Dodos*, to win the debate between evolution and intelligent design, scientists must battle public relations firms that understand "the need to tell simple clean stories, not constrained by the truth ... public relations firms that figured out the need for simple slogans ... instead of wasting time explaining entire stories to the general public."

Many scientists believe that, by not becoming popularizers of science, they somehow preserve their authority, or perhaps their dignity. However, says Stanford climatologist Stephen Schneider, "if we do avoid the public arena entirely, then we merely abdicate the popularization to someone else—someone who is probably less knowledgeable or responsible.... In my view, staying out of the fray is not taking the 'high ground'; it is just passing the buck."

There are benefits to being a public scientist—broad influence, for example—but there can also be costs. Carl Sagan offers an example of both. His renown brought him Emmys, a Pulitzer Prize, and considerable public influence for the issues that he valued—the primacy of science and of preventing nuclear holocaust. But it also cost him tenure at Harvard and membership in the National Academy of Sciences.

Another cost of being a public scientist is the uncomfortable need to walk an ethical tightrope, as described by Schneider in a 1989 interview in *Discover* magazine:

> On the one hand, as scientists we are ethically bound to the scientific method, in effect promising to tell the truth, the whole truth, and nothing but—which means that we must include all the doubts, the caveats, the ifs, ands, and buts. On the other hand, we are not just scientists but human beings as well. And like most people we'd like to see the world a better place, which in this context translates into our working to reduce the risk of potentially disastrous climatic change. To do that we need to get some broadbased support, to capture the public's

> imagination. That, of course, entails getting loads of media coverage. So we have to offer up scary scenarios, make simplified, dramatic statements, and make little mention of any doubts we might have. This 'double ethical bind' we frequently find ourselves in cannot be solved by any formula. Each of us has to decide what the right balance is between being effective and being honest. I hope that means being both.

Public scientists also must resign themselves to being misquoted—and even attacked for those misquotes—as is illustrated by how Schneider's quote above was, in fact, egregiously misquoted. In publishing an editorial attack on Schneider, the *Detroit News* truncated the quote, changing its point. The editorial deleted "I hope that means being both," and used the truncated quote to charge that Schneider was "prepared to play fast and loose not only with the truth but with the public psyche." In another attack on Schneider, business professor Julian Simon even *added* to the quote the words "Scientists should consider stretching the truth..." to bolster his charge that Schneider exaggerated global warming forecasts.

Public scientists must also commit to protecting themselves by extensively documenting their positions. Says Schneider in his online essay "Mediarology":

> Those who make public statements should also produce a hierarchy of backup products ranging from op-ed pieces...to longer popular articles..., which provide more depth, to full length books, which meticulously distinguish the aspects of an issue that are well understood from those that are more speculative. Books should also provide an account of how one's views have changed as the scientific evidence has changed. Even if only a minute segment of the public really wants this level of detail, this hierarchy of articles and books in the popular and scientific literature gives a scientist credibility in the popularization process.

Public scientists evolve in three basic ways: To reengineer a line from Shakespeare, some are *born* publicists, some *achieve* publicity, and some have publicity *thrust upon them*. In deciding whether you will become a public scientist, it is helpful to understand this evolution. For example, Carl Sagan was a *born* publicist, says Keay Davidson, author of *Carl Sagan, a Life*. "I think he wanted to be a public figure, not just a scientist, but he wanted to be a political activist; he wanted to be a celebrated author. He did all those things wonderfully, but he did pay a price for it."

Those who *achieve* the status of public scientists commit major time and energy—including writing popular books, articles, and essays, cultivating reporters, appearing on TV talk shows, testifying before Congress, and producing television specials, such as Sagan's memorable *Cosmos*.

And those scientists who have publicity *thrust upon them* find that it occurs in the most unexpected ways, as did Schneider. He recalls in his book *Global Warming* the moment he became "famous." The occasion was a talk at the 1972 AAAS meeting on human impact on climate. In an offhanded quip, he said, "Nowadays, everybody is doing something about the weather, but nobody is talking about it." Schneider recalls the unintended consequence of that quip:

> At the front of the audience a distinguished-looking gentleman was taking notes: he turned out to be the dean of all science writers, Walter Sullivan of the *New York Times*. Since journalists love one-liners—especially if they boil down complicated issues into a quick phrase or create controversy, the next day's *New York Times* featured a story on weather control that closed with my reverse Mark Twain quip. From then on, for better and for worse, my opinions were no longer my own property.

Shortly afterward, Schneider got his first taste of the downside of being publicly quoted. When he returned to his office at the National Center for Atmospheric Research, he found the *New York Times* article prominently posted, with the anonymous editorial comment "bullshit" stamped beside it.

Biologist JoAnn Burkholder, discussed in chapter 25, represents another example of a scientist who had publicity thrust upon them. However, rather than an offhanded quip, it was Burkholder's scientific findings of the toxicity of the marine algae *Pfiesteria* that generated publicity—as well as attempted suppression and persecution by a funding agency, politicians, and industry. In such a case, it is the scientist's integrity that requires him or her to accept the role of a public scientist. Burkholder advises scientists on how to cope with such controversial discoveries in her essay "Uncertain Ground: The Boundary between Science and the Press":

> There is no way to "plan ahead" and stay comfortably hidden in a laboratory away from the public eye; no "safe zone" to protect a scientist from stumbling upon data that completely change his/her life; no way to know when difficult ethical choices will arise. A scientist, as any person, is in training throughout his/her life to confront such decisions....
>
> In my view, honest communication with journalists and others in the general public is a scientist's professional responsibility. Whether in the laboratory or communicating with journalists and the general public, the scientists who act with integrity in even the smallest of daily

> decisions will have the best training to confront the challenges of the most difficult, often-highly publicized ethical choices.

Of course, being a public scientist need not necessarily mean advocacy and controversy. There are public scientists, such as physicist Michio Kaku, psychologist Steven Pinker, astrophysicist Neil deGrasse Tyson, and engineer Henry Petroski, who have enjoyed rewarding careers as public educators. In Petroski's case, he started with op-eds in major newspapers and went on to write engaging, enlightening books such as *To Engineer Is Human*, *The Evolution of Useful Things*, and *The Pencil*. Through those books, he has become known by journalists as an authoritative, straightforward source of clear information on engineering.

Science communicator Rick Borchelt might well be describing such public educators as Kaku, Pinker, Tyson, or Petroski when he says, "Reporters aren't looking for Carl Sagan; reporters are really looking for someone who is more low-key, but who can explain what they are doing but not lecture on a pulpit to them."

In the end, regardless of whether you decide to be a public scientist, your lay-level communications are invaluable to society. Every time you explain your work—no matter how modest the audience—you contribute to better understanding of science and technology. You help bridge the gulf between laboratory researchers with their critical knowledge, and the public with its need for that knowledge. For one thing, explaining science and technology offers the public a deep intellectual satisfaction, wrote Columbia University physicist Brian Greene in the *New York Times*:

> Like a life without music, art or literature, a life without science is bereft of something that gives experience a rich and otherwise inaccessible dimension.... Science is the greatest of all adventure stories, one that's been unfolding for thousands of years as we have sought to understand ourselves and our surroundings. Science needs to be taught to the young and communicated to the mature in a manner that captures this drama. We must embark on a cultural shift that places science in its rightful place alongside music, art and literature as an indispensable part of what makes life worth living.

Reaching across that gulf is also critical for society to meet the challenges of new science and technology, wrote Greene:

> When we look at the wealth of opportunities hovering on the horizon—stem cells, genomic sequencing, personalized medicine, longevity research, nanoscience, brain-machine interface, quantum computers, space technology—we realize how crucial it is to cultivate a general

> public that can engage with scientific issues; there's simply no other way that as a society we will be prepared to make informed decisions on a range of issues that will shape the future.

Sadly, the information gulf between scientists and the public is growing, both because of the dwindling cadre of experienced science writers and because of shrinking general media coverage of science and technology. Certainly, the science-oriented public has access to excellent information sources about science and technology—from such television shows as *NOVA* and the specialized science cable channels, to the many science magazines and books. However, science and technology have all but disappeared from the nightly news shows, daily newspapers, and other general media that act as a doorway into science and technology for the general public. Recall the "State of the News Media 2008" report of the Project for Excellence in Journalism, discussed in the introduction, which found that newspapers and network TV news devote only 2 percent of their coverage to science and technology. This absence of coverage can have serious consequences. For example, it can make it easier for politicians to marginalize scientific information in public debate. A classic case of such marginalization is the NASA public affairs office's notorious suppression of climatologist James Hansen's work on global warming during the Bush administration—discussed in the online working with public information officers section at *ExplainingResearch.com*.

Failure to bridge the information gulf between researchers and the public will hamper, perhaps tragically, our ability to solve the massive global problems we face—climate change, resource depletion, ecological damage, food security, and disease. As Sagan warned in an interview the year before his death in 1996, "We have…arranged things so that almost no one understands science and technology. This is a prescription for disaster. We might get away with it for a while, but sooner or later this combustible mixture of ignorance and power is going to blow up in our faces."

INDEX